The Fortunate Biped

A Very Brief History of Human Evolution

Michael Polley

Copyright © 2015 Michael Polley

All rights reserved, including the right to reproduce this book, or portions thereof in any form. No part of this text may be reproduced, transmitted, downloaded, decompiled, reverse engineered, or stored, in any form or introduced into any information storage and retrieval system, in any form or by any means, whether electronic or mechanical without the express written permission of the author.

.

ISBN: 978-1-326-29430-4

PublishNation, London
www.publishnation.co.uk

"We are just an advanced breed of monkey on a minor planet of a very average star"

Stephen Hawking

1

From Big Bang to Sliced Bread

Unless you are an employee of NASA, or your name is maybe Stephen Hawking or God, you might be more than just a little surprised to learn that somewhere around 14 billion years ago our universe was actually nothing like the universe we know today. There were no stars, no planets, no dark matter, no black holes, no infinite expanses of dark space, no Earth, no you or me. No, at that time the universe was actually quite the opposite. It existed in a state where there was no concept of space or even time, and everything, absolutely everything, was compressed into one single infinitely small, infinitely dense dot that just, well, was. In fact, that's all there was.

Hard to imagine I know. But imagine throwing everything you own, your house, your car, family, friends, possessions, even your mother-in-law, everything in your "universe" into one giant suitcase. Now imagine that some clever science bod beavering away in the bowels of the local university had a machine with which he could shrink that suitcase into a smaller and smaller space until it was the size of just a tiny atom. Now ask him to sharpen his pencil and straighten up his lab coat, and try to shrink that tiny atom even further until it was less than a billionth of its new atomic size.

Now try to imagine the same thing happening but with everything in the whole universe, squeezing absolutely everything, every star, every planet, every galaxy, every single mother-in-law, all into one infinitely small object, something a billion times smaller and infinitely more dense than a tiny atom. This then was the universe as it existed approximately 14 billion years ago. Actually, at this point it wasn't a "Universe" at all, in fact it was the complete opposite, and it's what clever science bods call a "Singularity".

Presumably things had quite happily existed that way for billions of years (even though it seems there was no time to measure it by) and there was no expectation that anything would be different any time soon. Things were good in their own, if somewhat cramped, way. Then, all of a sudden, for no apparent reason all hell broke loose and that infinitely small and dense dot exploded, and not just any explosion, it was an explosion of such magnitude that it threw everything that was compressed into that tiny over-cramped Singularity in all directions and with such force that it ultimately turned our infinitely small dot into our infinitely large Universe.

Now we, or more importantly the physics boffins who have IQ's higher than most of us can count, don't know all this for sure. They believe it to be true because it is the best explanation their collective sizable brains could cook up and agree upon, and if indeed this Singularity did exist they have no real explanation as to what might have caused it to then suddenly explode in the first place. Maybe our little Singularity actually had an internal "Countdown to Launch" clock all along that just finally reached zero, maybe it just had a hissy fit, maybe it just couldn't hold in all that tightly packed "luggage" any longer, or maybe it just got bored with its extremely cramped existence and decided to stretch out a little, likely neither we nor the boffins will ever know. But whatever the reason, our little Singularity did indeed blow, and the result was our universe.

In time, one particularly clever Space boffin named Edwin Hubble, went on to do some clever things with numbers and a slide-rule, and concluded that the force of this explosion was so great that apparently even today the universe is still continuing to expand and still forcing matter outward from its initial source. Further work on the chalkboard by other highly paid number crunchers calculated the speed of this expansion, and then with some additional overly clever science stuff and even more complex number crunching, they eventually worked out that our tiny Singularity must have exploded somewhere around 13.7 billion years ago.

That one single event almost 14 billion years ago that created our infinitely vast universe was not just a momentous event, it was "the" momentous event, the one single event around which everything that we know, everything that has and will happen, everything that exists, all originated from. So actually quite important then, and deemed so important by the guys in lab coats that they gave it its own (and given its significance rather unimaginative) name, "The Big Bang", and hence our universe as we now know it was born.

However, while we waited for today's boffins to grow brains large enough to even imagine a "Big Bang" theory, many earlier theories had been put forward to explain the creation of the universe. Early civilizations that had little knowledge on which to base their theories other than a daily view of the sun rising and setting and a night sky full of twinkling objects, believed that their universe was created by strange mythical creatures such as water serpents, a universal Mother, or even hatched from a cosmic egg (although there seems to have been little thought or concern as to what may have laid this cosmic egg in the first place).

Other later theories include what's known as the "Goldfish Bowl" theory which is premised on a belief that the universe was actually constructed in an Alien Laboratory, where some super-intelligent alien species, presumably bored and with a large amount of time and money on their hands, are merely conducting a massive galactic science experiment. There is also what's known as the "Matrix" theory, which surmises that the universe is actually a massive computer simulation driven by a "self-aware" Super computer designed to keep us all in a happy dream state while reality has us all actually hooked up to the machine feeding its needs.

Theories with slightly less "Hollywood" and a little more "science" have also been presented offering an alternative to the Big Bang. These include the "Bubble" theory which has our universe as just one of multiple universes, or the "Steady-state" theory which rather philosophically states that the universe actually has no beginning and has no end, it has always existed and will always continue to exist, and thus conveniently removing the need to spend countless hours wondering how the universe was formed in the first place, it's just always been there.

Then of course there is the theory of "Divine Providence", sometimes called the "Intelligent design" theory, and is a belief that the universe and everything in it is the masterpiece of an old man with a large grey beard, wearing a long white gown, and sporting a comfortable line in sling-back sandals. Scriptures report that He managed the remarkable feat of first creating the whole universe and then going on to make mankind in his own image (although in this particular respect he may have over-estimated His ability), all in roughly 6 days, leaving himself the rest of the week to relax, sit back, and admire his handy work (although, some do say that on the seventh day just for His own amusement He made Taxmen, Lawyers, and Politicians, just to annoy us all).

Part of the reason this belief continues to hold support is that strangely it is not necessarily contradictory to the Big Bang theory. Religious doctrines still imply the universe was created in one single (albeit six day) act, the key difference being that they have the finger of "God" on a button that said "Push to initiate creation sequence". To them the creation of the universe was a conscious and planned decision, of course from a scientific viewpoint this is also not a million miles away from the aforementioned "Goldfish Bowl" alien experiment theory.

Now those overly bright Physics boffins continue to try to reconcile their own Universe and Big-bang theories in terms of complex mathematical formula, and some rather clever "you need a brain the size of Neptune to understand it" stuff they call quantum mechanics and string theories, but in reality they have no real proof. It's just complex physics "mumbo-jumbo" that give them a possible version of events that stand up to scrutiny against what we (or more accurately they) do actually empirically know about the universe and the mechanics by which it works today. But just like the rest of us who aren't blessed with the pre-requisite Neptune-sized brains, they really have no idea (and most importantly no way to prove) one way or the other what actually did happen before, during, or after the Big bang. However, their theory of "Singularity plus Big Bang equals Universe" seems to help them all sleep comfortably at night, and so not wanting to argue with people whose brains are clearly far larger than mine, it's a theory we should all just accept as probably correct, at least for now.

As for our own little Solar system, it is estimated to have been born about 9 billion years after the Big Bang, making it about 4.7 billion years old. According to current estimates, our Sun is apparently one of more than 100 billion stars in our Milky Way galaxy (although the man tasked with counting is apparently still counting), and orbits roughly 25,000 light-years from its galactic centre. Our Solar system consists of the Sun, (that big bright thing in the sky), eight official planets and at least three "dwarf" planets (wannabe planets who it seems are just not quite big enough to make the planetary cut) such as Pluto, which itself was rather ignominiously down-graded from full membership in the exclusive Planetary Club to "dwarf planet" status in 2006.

To help visualize the relative size and distances of these objects in our Solar system imagine a model where everything is reduced in size by a factor of one billion. In this model the Earth would be about the size of a grape, and the moon would be peanut-sized about 30cm away from it. The Sun would be a ball about the size of a grown man across and about 150 metres from Earth. Jupiter would be the size of a grapefruit and about half a mile away from the Sun, while Saturn the size of an orange and a mile away from the Sun. Finally, Uranus and Neptune, our solar system's outermost planets, would both be about the size of an apple, and about 2 and 3 miles away from the sun respectively. To provide further perspective, under this scale, a human being would be about the size of an atom.

As for our little planet, Earth, it is the third planet from the Sun, sitting quite comfortably about 93 million miles away from it and has an orbital speed around it of about 30km per second, (which is fast enough to get a man to the moon in about 3 hours), and thus completes that orbit once every 365.25 days. Roughly translated this means that the Earth travels through space at about 66,600 miles per hour, which is about eight times faster than a speeding bullet. Our Earth also has a circumference of about 40,000km making it the fifth largest planet in the Solar system, and is also apparently the only planet that currently supports life (at least as we know it, Jim).

Earth is also the largest of what planetary-boffins call the terrestrial, or rocky, planets (the current club members being Mercury, Venus, Earth and Mars), which not unsurprisingly then are primarily composed of rock, and have a relatively high density, slow rotations around the Sun, and solid surfaces (which is rather handy for those of us who choose to live on, rather than in, their planetary home). In comparison, the remaining four planets (Jupiter, Saturn, Uranus and Neptune), are primarily composed of hydrogen and helium and thus are collectively known as the "Gas planets".

Rather disappointingly, the Earth is also the only planet in our Solar system whose name does not derive from either Greek or Roman mythology. You would have thought that the people entrusted to sit around a table at Galactic HQ deciding the names of planets would have at least saved a selection of the best mythological names for our own. Planet Zeus, or Hercules, or even Spartacus, would have maybe been better options than settling for a name that is effectively just a synonym for dirt. But it is on this our rocky planet, now known officially throughout the universe as planet Earth, found in a small unassuming Solar system within the Milky Way galaxy, that our unique evolutionary story of mankind will unfold.

In reality though, for the likes of you and me who already struggle to understand how many of our life's day to day phenomenon actually work (such how a huge passenger airliner carrying over 300 people manages to get off the ground and stay off the ground), is that outside of the odd game of Trivial Pursuit an understanding of the theoretical physics behind the creation of the universe is unlikely to benefit any daily task or conversation. What I do need on a day to day basis is a practical understanding of how to pilot my car from home to work, at least a rudimentary understanding of child and female psychology such that I can avoid raising a drug addicted psychopath and divorce, a basic knowledge of math to avoid being short-changed at the local Supermarket, and just enough skills at work to keep me one step ahead of the guy who's angling for my job.

Maybe then we don't all need to know the exact mechanics of it all. Maybe the true challenge for our High School Science teachers is not just to teach spotty faced teenagers that the Earth is 93 million miles away from the Sun, but is actually to inspire in them a sense of wonder at the beauty of our universe, and maybe more importantly a sense of respect for our place in it. Because for what started as that infinitely small, infinitely dense dot which eventually became our infinitely large universe, filled with a constantly changing canvas of trillions of heavenly bodies that fill the infinite void, all seemingly moving in perfect harmony and balance, truly is a thing of sheer beauty. The universe is seemingly random and unthinking, it is mechanical in nature and without purpose, and yet somehow it all just works, and does so in a way that results in the singularly most beautiful creation we could possibly imagine. Best not to question it, just enjoy it.

However, what is important to our story here is that in that one single cataclysmic "big bang" event we see set in motion a series of events that, over the course of the following 13 billion years or so, eventually led to the creation of stars with giant balls of rock circling around them, and where on one of those giant balls in a far flung corner of the Universe, tiny life forms would manage to grow and crawl out of some rather toxic, and probably quite smelly, primeval sludge. Those tiny life forms would then miraculously go on to evolve over another rather lengthy period of time to a point where that tiny planet would eventually give rise to a rather special species, a species who would give themselves the rather catchy name of *Homo sapiens,* and also name the large ball of rock they call home after some brown stuff under their feet. A species that would learn to control fire, invent the wheel, build some rather useful tools out of stone, and create numbers with clever ways to put them together, such that they would eventually help explain why there was a Big Bang in the first place.

Indeed, it seems that from the very first instance of that giant explosion, the way it exploded, the speed with which it projected matter outwards, the direction in which it projected that matter out, every single seemingly random event that happened in that first split second, actually seems to have ensured that all the necessary ingredients needed for a tiny bag of chemicals to miraculously spring forth into life, would all be fully present and accounted for in the toxic melting pot that was to form on one (or more) of the universe's infant planets, and from there, through the miracle of evolution and a healthy dose of luck, setting in motion a chain of events such that some several billion years later, Mr. and Mrs. H. Sapien would eventually walk upon planet Earth.

However, it's important to note that even though the Universe had seemingly conspired to conveniently create all the necessary ingredients for life to emerge on our tiny ball of rock, our own illustrious presence on the planet was by no means an evolutionary foregone conclusion, evolution and luck still needed to play their part in our story. It seems that up to about 7 million years or so ago we (or rather our somewhat hairier ancestors) were still perfectly happy just playing around in their home up in the trees of East Africa, eating bananas, picking fleas from each other's hair, and generally enjoying life as fully fledged members of the chimpanzee family tree. Then it seems evolution took the rather radical step of taking what had been to that point a perfectly functional and happy evolutionary line and split it in two, and it did so when one bright morning in that distant past, one of those early ancestors of ours woke up and decided that they were bored with just climbing around in the trees all day, and thought they would give walking around down on the ground a go, and that they would do so upright on two legs rather than knuckle-walking around on all fours like a French rugby player.

Happily it seems there was no passing sabre-toothed tiger or other starving predator passing close by as that first ape-like ancestor of ours took his first nervous steps onto the savannah (else evolution's glorious experiment in mankind may well have ended before it began), and soon others followed the trend, thus leaving our chimp cousins behind up in the trees and so starting our own unique evolutionary line that walked around on the ground comfortably stood upright on two legs. All that was left then was to wait for another bright spark in our new ancestral line to somehow figure out how to control fire, then another to have the bright idea to use stone to fashion tools, and we were well on our way to walking on the moon and sliced bread.

There were however also a lot of other seemingly unrelated but remarkably fortuitous events that worked in our favour (the giant meteor hit that caused the extinction of the dinosaurs), near misses (the Neanderthals), and lucky breaks (if the Earth were a mere 1 mile closer or further from the sun, we would all instantly either fry or freeze respectively), that were also needed to ensure we did eventually meet our destiny. But at that single 1 second past Big Bang time, it seems the scene had indeed been perfectly set for life to emerge, leaving evolution and luck to work their magic such that a species such as ours would ultimately be destined to evolve somewhere in the Universe, and to ultimately evolve into the intellectual masters of their planet (at least in the opinion of all the animals who returned the questionnaire).

Bearded scholars in knitted sweaters tell us that apparently there were likely six key evolutionary adaptations that seem to have been crucial to our unique success as a species. Our intelligence coupled with its ability for abstract thought, our linguistic communication, a highly developed eye (well two actually), hands with opposable thumbs and an ability for fine motor movement, the afore mentioned ability to walk upright on two legs, and finally our highly social nature, (a controversial seventh key adaptation, an ability to willingly devote over two thirds of our lives to spending at least 5 days a week working for someone else, doing something we don't particularly enjoy, all for very little reward, has yet to receive formal recognition from those same bearded scholars).

True, other species possess potentially comparable intelligence such as dolphins, birds of prey have eyesight to rival or even exceed that of our own, the great apes have hands with opposable thumbs (although not as dextrous as that of the human hand), but no other single species possess more than just one or two of these key adaptations. *Homo sapiens* have uniquely evolved to possess all six. We are it seems the state-of-the-art, fully-loaded, concept car of natural evolution. Indeed we seem so remarkably suited as a species to life on planet Earth that as a species we still spend a great deal of time debating whether we really could have been that lucky and evolved merely as just an incredible fluke of pure evolutionary chance, or whether we were indeed really just all part of some great universal master plan orchestrated by a divine being in a flowing white robe.

This conflict between Science and religion (Creationists versus Evolutionists) still continues to take up way too much valuable time of many of today's great minds, time probably better served solving slightly more pressing issues such as world hunger, global warming, and why pairs of socks go into a washing machine but only single socks emerge. Despite the increasing weight of scientific evidence to support the theory that we are indeed just an incredibly lucky benefactor of nature's evolutionary process, the debate will likely continue right up to the point where fate steps in and while we're all still busy arguing, diverts a moon sized asteroid towards Earth making the whole discussion moot.

However, the uncomfortable truth (for some anyway) seems to be that as a species we are indeed merely just the result of exactly the same evolutionary forces, the same natural selection, and the same survival of the fittest crap-shoot as all other life forms on planet Earth (or for that matter anywhere else life that may exist in the universe). In this respect we can probably be viewed as just well-behaved clean-shaven monkeys and no different to the dinosaurs wandering around in their global Jurassic Park some 70 million years ago living out Nature's then evolutionary experiment around "biggest is best". The key difference in our favour being that we had the good fortune to have evolution wait until after a meteor the size of Texas had slammed into the Earth effectively wiping out almost all its evolutionary work to date, including the dinosaurs, before settling down to start work on our particular evolutionary line.

So this then is a story about the evolution of a species, a very fortunate species, and how it ultimately was to emerge on one tiny planet in a Solar system in a far flung corner of the universe. It's a story of how a suitcase that contained all the essential ingredients needed to create a universe that was somehow all squeezed into an infinitely small dot, eventually exploded, and in doing so set us on that path towards Mozart, space travel, and sliced bread. It is part mystery (just how did all this happen?), its part scientific fact (the boffins can explain some of it), and its part science fiction (the boffins can really only guess at most of it). It's a story with heroic characters such as Charles Darwin, Louis and Mary Leakey, and the first man who decided to risk eyebrows and singed fingers to see just what fire was all about. It's a story that will cover billions of years, and has its beginning rooted in the stars and ends in the infinite possibilities of the human mind. It's the story of you and me.

It's also a story about perspective. The universe itself is immense with billions of Galaxies, each with trillions of stars, and each with its own planets to orbit them, our tiny Earth is just one of those trillions and trillions of planets to be found in our universe. As incredible as the evolution and subsequent achievements of mankind have been, when compared to the universe's many other wondrous events outside of our planet's own little sub-plot they are somewhat insignificant in comparison, and certainly insignificant to the miracle of the universe itself.

We have only existed for a very small fraction of the 14 billion years or so the universe has been creating galaxies, stars and planets, and likely there have been many other incredible little "miracles" elsewhere in space that rival our own, but the universe has always moved on. Looking up at the night sky full of stars we are looking at a limitless unfolding story on an unimaginable scale, and in each little speck of that vista a story in miniature is likely unfolding. This is just one of those trillions of stories, but for you and I at least, still a very important one.

Would a moon-sized meteor crashing into Earth put something of a roadblock in the path of mankind's evolutionary march, yes it probably would. Would it make the universe wobble and change the nature of its existence, it would not. Thus, it's also a story about gaining an appreciation for mankind's place in the universe and in the overall picture of life on planet Earth, an appreciation that we are only part of a much greater whole.

But mostly it's a story that tries to tell just one possible version of events in which, to loosely paraphrase Mozart, we have all the possible theories we just need to assemble them together into something that makes sense. It's a story which leads us to mankind's existence and down a path to such remarkable achievements as walking on the moon, the Mona Lisa, the Ferrari Enzo, the push-up bra, and sliced bread, but similarly to some unmitigated disasters such as the helicopter ejection seat, Motorola's battery powered battery charger, the solar powered torchlight, and spandex.

However, the goal of this book is not to be a textbook, it is not to be mistaken for a study guide, scientific document, or source material. Use the contents of this book as a reference for homework or a science project and you do so at your own peril and in almost certainty of a failing grade. The management acknowledges that any of the facts stated within these covers may well be publically declared by the world's collective boffins as either being overly simplistic, misinterpreted, or actually in some cases possibly even down-right wrong. Everything in this book then should likely be read with the accompaniment of about 2 pounds of sea-salt, it is after all just a story.

But as we shall see even the boffins themselves with their watermelon-sized brains, chalk boards, and calculators, still do not have all the answers to just how that first tiny spark of life that popped its head out of that primordial sludge some 4 billion years ago actually came to be, or how it then went on to rather fortuitously result in a somewhat mysterious biped who would prove to be evolution's crowning achievement (at least on planet Spartacus). In the murky world inhabited by those who seek to unravel the mysteries of human evolution the answers to the questions being posed would seem to be harder to find than Osama Bin Laden, and so our understanding of our own unique evolutionary story has really only started to form in the last 100 years or so (with as we shall see, several hiccups along the way), and even now there remains much debate around the course that evolution actually took such that you would be here today reading this book. All of which means that I'm probably on pretty solid ground academically anyway regardless of what appears in the coming pages, likely there's some science-type out there somewhere who will support the facts and theories herein (at least for a suitable sum of money anyway).

Maybe at some point in the future the boffins will finally discover the true nature of mankind's evolutionary story such that we can have one universally agreed version of the truth. Hopefully it won't be something ultimately only explainable through volumes of complex scientific theories that all just look like the random squiggles of monkeys to most of us who are unfortunately only blessed with just a few IQ points north of goldfish.

But until then, what follows is probably as good an account as any….. I think.

2

Out Of The Primordial Soup

When the world first read how Mary Shelley's fictional eccentric scientist, Victor Frankenstein, had created life by passing enough electricity to jump start a Jumbo jet through his rather large patchwork cadaver, she was not only creating a gothic literary masterpiece to scare the pants of most readers, but was also indirectly introducing the world to the notion that maybe life could indeed be created from some readily available bits and bobs, and a little scientific jiggery-pokery. Indeed, a quick canvas of the average factory floor will still indicate a persisting belief that passing something like the full force of the National Grid thru a lifeless assembly of inorganic materials will result in it miraculously sparking to life (bolts through the neck optional).

Now this may just be more of a testament to the strength Mary Shelly's literary skills on the psyche of the public, but it does also speak to how little the average person knows about the creation of life, the forms it initially took, and the circumstances required to make it all happen in the first place.

The creation of life on Earth, or actually anywhere else for that matter, is matched only by the creation of the Universe itself in respect of its significance, and much like the debates surrounding that other great event, the creation of life has been the subject of numerous theories throughout the ages as to its possible origins. Some of these theories are natural progressions of those presented earlier to explain the creation of the Universe itself, and thus come as part of a convenient one-stop package deal. By accepting the solution for the creation of the Universe, you get the low-down on the creation of life thrown in for free. Buy one, get one free.

Such theories would include the previously mentioned "Goldfish Bowl" theory, where one would imagine that if as part of a grand alien experiment some extra-terrestrial Lab jockey had actually managed to create the Universe, he would naturally have gone on to populate his Universe with some life forms, at the very least to provide some variation and interest. As all good students know, going that extra step can be the difference between a simple passing-grade versus a gold star.

The other obvious example would be the results of the well documented week long endeavours of an all-seeing, all knowing Universal Director of Operations, whose efforts in the first few days concentrated on creating the Universe itself, but whose particular work on the sixth day apparently resulted in both Adam, Eve, an apple, an asp, and all other "living creatures".

Still other beliefs with a similar religious basis (but from the more easterly side of the world map), would have us all caught in an endless cycle of birth, death, and then rebirth, with even some variations having you reborn each time as one of any number of creatures depending upon your behaviour in your preceding physical incarnation. A life lived full of caring, sharing, tree-hugging, listening to folk music, and general philanthropic endeavors, would see you seemingly reborn in human form or even raised to saintly status, while one spent neglecting such noble goals would see you reborn as a more "lowly creature" as befitting your apparent deviation from such righteous pursuits. To such religions, the Earth would thus seem to be populated with only human "souls" in various physical incarnations indicating past performance for all to see. As far removed as these ideas are from any scientific basis, it would at least offer an explanation for the proliferation of such reviled creatures as rats, snakes, mosquitoes, cockroaches, and mother-in-laws.

Another theory that has a slightly more scientific basis, and one that also has a degree of support from certain areas within the world of chalk boards and lab coats, is the theory that perhaps life did not begin on planet Earth at all, but that it was transported here from elsewhere in space, either from within our own Solar system, or maybe even from another star system in another galaxy (far, far, away,...), a theory known as "Panspermia".

For instance, rocks regularly get blasted out from Mars by cosmic impacts, and a number of Martian meteorites have indeed been found on Earth, meteorites that some researchers have controversially suggested brought along with them living microbes which subsequently colonized Earth and became the true catalyst for emergent life on Earth. A notion that if ever proven true would make us all in part at least descendants of Martians.

However, the major stumbling block to this particular theory of Panspermia is that it would require life to have actually existed on the "home" planet in the first place in order for it to have "hitch-hiked" to us on that conveniently directed rock (at the time of writing at least, I can confirm that all searches for any such "home" planet have so far drawn a blank, so for now, E.T would struggle to know which way to point his bike before cycling home). Our problem here would be that even if this concept of Panspermia were indeed to be true, the question of how life began on Earth would then only have shifted to be one of how life began elsewhere in space. The original question would remain unanswered, but at least our search for extra-terrestrial life would be over as it would have apparently been with us all the time.

However, for over two thousand years, the prevailing theory, one first put forward by the ancient Greeks, and one which surprisingly continued to have support with western scholars right thru to the 19th century, was a belief in what is known as "Spontaneous generation". The theory held as its central belief that certain living organisms could spontaneously result from other decaying organic substances, mice from rotting hay, frogs from slime, flies form putrid meat (and presumably lawyers, politicians, and accountants, from whatever's found on the bottom of your shoe).

The theory was only finally disproven by the experiments of a certain Louis Pasteur (1822-1895), a French chemist more familiar to us for his invention of a method to treat milk and wine in order to prevent it from causing sickness by a process that came to be called pasteurization (something for which mums and hobos the world over will be forever grateful). His experiments showed for instance that the likes of maggots only appeared on rotten meat when exposed to flies who could then lay eggs, but if left in a sealed container, preventing any access to any conveniently passing flies, no maggots (or anything else for that matter) were able to "spontaneously" appear.

The alternative to any such spontaneous generation of life, as debunked by Pasteur and his milk drinking scientific friends, seemed to be the theory of "biogenesis", a theory that holds at its foundation that every living thing came from a pre-existing living thing. By the middle of the 19th century, the theory of biogenesis had accumulated so much evidential support, that the earlier held notion of spontaneous generation had been effectively down-graded from an unquestioned theory to just an amusing dinner party "anecdote".

Now this was all well and good, but now this created a bit of a problem for those still wishing to explain how life actually began. If we are to believe that all living things came from other pre-existing living things, how then did the process start in the first place? We were effectively back to the same problem we faced with the theory of panspermia, and thus back to square one. The explanation for the creation of Life was starting to become a bit of an embarrassing hole in science's book of "101 things we really ought to know".

Then, in the 1920's, from behind the iron-curtain a certain Alexander Oparin proposed a possible solution to help save the blushes of the scientific fraternity. He proposed that the idea of the "spontaneous generation of life" that had earlier been refuted by Louis Pasteur, did in fact occur once during the infancy of the Earth. That early hostile and toxic atmosphere, a "primordial Soup" concocted out of the prevailing gases, radiation, electrical storms, sunlight and the downright primeval conditions on Earth at that time, could in fact have created the conditions needed to "spontaneously" create organic molecules which could then themselves combine in evermore complex ways until the necessary building blocks to "spontaneously generate" life (admittedly a very basis form of life, a minute single-cell organism rather than a fully formed Frankenstein's monster) emerged.

Oparin further reasoned that our atmosphere, which over time had happily become the oxygen-rich atmosphere more friendly to the complex life-forms we see today, now prevents such a synthesis, and thus such spontaneous generation of life was itself now impossible because the conditions needed to spark such emergent life existed only under those initial conditions found at the time of Earth's chaotic infancy.

Oparin proposed that once the spark of life had "spontaneously" emerged from Earth's early primordial soup some several billion years ago, the baton then needed to be handed over to the process of biogenesis (what we now know today as "evolution") to continue and complete the work. And so in a rare twist for the scientific bods, it seems that in part everyone was to some degree right, and so to this day many modern theories for the origin of life still happily take Oparin's ideas as their starting point.

So how then do we believe that life did miraculously, and spontaneously, emerge on the infant planet Earth? Well, to try and understand, we first have to go back to those first initial days of our planet's birth. Now the people who are paid to know these kind of things tell us that (give or take a day or two) the Earth was formed roughly 4.5 billion years ago from the gas and dust which formed around our newly created Sun, this matter aggregated under the force of gravity into clumps which then grew in size as they themselves then attracted more material, became bigger and bigger clumps, gradually creating a clump large enough such that science bods could safely call it a planet.

This sizable clump of matter, later to be named "the Earth" by the most advanced example of its soon to be kick-started evolutionary process, was initially molten due to the violence of its formation, and thus effectively being nothing more than a planet-sized ball of fire. Such enormous heat was being generated by the intense gravitational compression that was occurring as our fiery ball's interior became more and more compressed under the increasing weight of the additional material that was being drawn onto its surface, and then as that material finally came to rest and its energy of motion was converted to heat energy.

As its temperature increased and ultimately crossed the melting point of iron (roughly about a lip-chapping 1,500 degrees Celsius), the iron sank towards the center of the Earth in what is dramatically called the "iron catastrophe", and which was at least in part responsible for the formation of the Earth's iron rich core we know today. This "catastrophe" released even more heat, raising the temperature of the Earth at that time by another couple of thousand degrees or so. Clearly, if life was going to emerge from this planet of ours, it certainly wasn't about to happen in these over-heated conditions.

This all appears to have been a relatively rapid process, at least on a galactic scale, such that by about 60 to 80 million years after its formation, the Earth was nearly complete (it was big, it was round, and it was solid). Earth had achieved about 99% of its current mass, had a rich iron core, and water was present (admittedly though only in the form of vapor released from the heating of the earth's early material). As a planet, the Earth was now effectively up and running, and online.

However, just to complicate things a little it seems that the newly formed planet Earth was being used as daily target practice for numerous asteroids which were constantly bombarding its surface throughout this period as the Solar system, itself still newborn, started to settle down from the chaos of its own birth. In keeping with the dramatic and grandiose names given to this period of Earth's history, this particular period has been named the time of "the Great Bombardment". Even if life could have emerged at this time on our Solar system's latest planetary addition, it likely would have been quickly squashed out (quite literally) by the daily blitz of incoming asteroids.

By the time the Earth was 100 million years old an early atmosphere had formed, it was however not what we would recognize as our atmosphere of today, in fact it was far from it. The high temperatures had led to out-gassing from newly formed rocks which combined with volcanic outpouring produced an atmosphere totally void of any oxygen, and instead one high in methane gases with some hydrogen, nitrogen, water vapor, and carbon dioxide, all thrown in just to spice up the toxic mix. Exposure to Earth's atmosphere at this time would see you likely live just long enough to fire off a few choice expletives before experiencing first-hand the type of death we imagine astronauts would face should they inadvertently crack the visor of their space-helmet while on a space-walk.

The atmosphere was also a great deal denser than our fragile bodies are used to moving around in too, with a surface pressure of around 250 atmospheres, compared to our comfortable 1 atmosphere of pressure at sea-level that we enjoy today. To give you some perspective, this is the sort of pressure you would feel pushing on you should you suddenly find yourself around 2,500m under the ocean, which translates to the equivalent of you trying to support the weight of around 50 Jumbo Jets, as such, you would very quickly find yourself one dimension short of your normal three-dimensional form.

And so, stepping out onto the surface of our planet at this time in its history it's hard to decide whether you would have first been crushed by the enormous atmospheric pressure, dissolved by the noxious vapors, hit by an asteroid, or whether you would just have suffocated from the complete absence of anything even remotely oxygen-like to breathe. So at this point then, still not the most appealing of places to visit let alone be the chosen site for any spontaneous generation of life. So at this point, life was still looking decidedly unlikely on the newly formed planet Earth.

Eventually though the Earth thankfully started to cool down to something vaguely bearable, its crust started to form and stabilize, and liquid water started to exist on the surface forming great oceans. Although now a little closer to something we might recognize, this was still far from any vision of a world where waves gently lap onto a white sandy beach under a perfect blue sky. No, these newly formed oceans were still green and highly acidic from all the dissolved iron and minerals, and the sky was orange from the high concentrations of methane, there were constantly erupting volcanoes, and little sunlight penetrated through to light the surface. This was no still place for any idyllic vacation getaway.

This period in the Earth's history is known to those who like to give names to large chunks of time as the Achaean period, and rather surprisingly it was actually during this time, some 3.8 billion years ago, that David Attenborough-types tell us the first semblance of life is believed to have crawled out of the primordial sludge. Actually, whatever life did indeed first emerge from this Achaean world didn't crawl anywhere, it didn't walk, talk, swim, scratch its bum, or do very much of anything at all, it was likely nothing more than a very simple bacteria, but nonetheless it was a form of life, it absorbed nutrients for energy, it could reproduce itself, and it fidgeted around a bit. It was as basic as life can get, but it was life, and from this very simple collection of twitching chemicals, all life on earth eventually evolved.

Dedicated science bods (who seemingly spend their entire waking hours agonizing over these kind of things), actually now believe that this spontaneous creation of life was merely a logical event, not a chance occurrence, but the inevitable bi-product of the conditions that existed on Earth all those years ago. Life it seems may well have been just a natural consequence of that exact same toxic atmosphere, acidic oceans, volcanically charged "hell on earth" (literally) that would prove so deadly to all of today's living organisms. It seems that 3.8 billion years ago the Universe had somehow created conditions on our tiny little planet that meant life was an unavoidable happy consequence. Indeed, if we are to believe our science bods, and clearly they know more about this stuff than we do, it seems life could only have begun in the presence of such conditions, and right on cue, it promptly did.

The newly formed shallow seas, so seemingly toxic, yet so rich in chemicals, created that "primordial soup" which when combined with volcanic activity and electrical storms, created the conditions for the first emergence of life to spring forth. It only needed to happen once, but once it did it set in motion the extraordinary chain of events that some 4 billion years later led to mankind, and ultimately to you being here right now gazing over the pages of this book.

If we are to believe then that given the right conditions and sufficient luck, life can, and will, emerge, it would clearly beg the question of whether we should expect that life would similarly emerge on other planets given similar conditions? Regardless of where and when, the basic building blocks and the conditions needed to spark them to life, presumably remain the same (the recipe for apple pie is the same whether you bake it in Australia or Alaska). Any newly formed planet roughly the size of the Earth, orbiting a star similar to that of our own sun from a distance of approximately 93 million miles, would presumably evolve under very similar forces and conditions as those experienced by our own infant planet. And as we are reliably informed, under such conditions life it seems is almost inevitable.

Thus, given a Universe filled to the brim with billions of galaxies with all their associated stars and planets, is it really such a stretch of the imagination that other similarly orbiting planets have evolved with a very similar early atmosphere, intense volcanic activity, and resulting oceans made of the necessary "primordial soup"? The

Universe it seems has a cookbook with a recipe for emergent life, it just needs a suitable and fully equipped kitchen in which to rustle-up the meal.

So emergent life could only happen the once on our own little planet (assuming we are all still going with Comrade Oparin's theory) at the single point where all the stars of an early toxic atmosphere, strong electrical and volcanic activity, and chemically-rich bodies of water, all conveniently aligned. However (and pay attention here as this bit is important) once that first basic bacterial life had popped its head out of the primordial soup, its subsequent survival and its potential path to more complex creatures such as you and I, was a journey that needed to be driven by the lottery of evolution and the vagaries of the Universe and all its potential to inflict unwelcome and cataclysmic events. Emergent life may well have been almost inevitable on our little ball of rock, but you and I, and every other life-form in-between, was not.

The Universe's cook book recipe for "Emergent life" would appear to be a one-time only event on planet Earth, but the recipe itself should be no one hit wonder, it is very likely an entirely repeatable dish on any other conveniently placed planet, but what nature then does with that initial little bundle of twitching life, if anything, is entirely down to the random whims of the Universe and evolutionary forces.

3

A Recipe for Life

Clearly, whatever it was that popped its head out of the primordial soup and that our science bods have joyously declared as "life" was no Frankenstein-type creation, it had no arms, legs, head, or tail, it couldn't grunt, groan, walk, see, feel, or moan at you because the washing-up is not done, it was as basic as life can get, but apparently the people who get to decide these things decided it was indeed "life". So just what made this little miracle "alive" as opposed to just a collection of chemicals throbbing away in the sludge? Where do we draw the line that says everything past this point is deemed to be a living organism?

Now you'd think that coming up with a definition for what constitutes life would be easy enough, but strangely it would appear that the collective brains of the scientific world cannot actually agree on a good definition of what actually is "life". It seems that what constitutes life (in the minds of the boffins at least) is a very abstract thing, and defining at what point we say something doesn't have the necessary components versus when it does is seemingly a very murky thing. The main problem with any definition for what may constitute life is that they tend to have loopholes.

Take for example the not unreasonable idea that "life" should be able to reproduce itself such that as a species it can continue to exist long into the future, on the face of it a fairly straightforward and quite logical premise (the only alternative being immortality, and as at last time of asking, it appears that the science bods still haven't quite cracked the key to this one yet). However, taken literally this definition would automatically exclude for example animal hybrids, such as the mule, which are born sterile, or more worryingly any man who was effectively firing blanks in the bedroom department, both of which by this definition would be deemed rather embarrassingly as "non-living". Confusingly, such a definition would also mean we would also need to include such things as fire, which can indeed reproduce itself, but (hopefully) quite logically is not itself a living thing.

Or how about the idea that a living thing should be able to metabolize, that is, take in energy to help it move, grow, and expel waste. But under that definition we would need to include such things as cars and planes, which can all "metabolize" fuel. It seems that life is such a complex thing, with so many components and moving parts, each of which are essential to its existence, that at the moment it seems just a little beyond today's great minds to agree a clear and definitive definition of just what constitutes a living organism.

What we are left with then is something of an uneasy compromise, a definition for "basic life forms" at least, that for now our own brainier specimens (themselves presumably deemed fully "alive") have begrudgingly agreed is as close as we can get pending some presumed future scientific (or mystical) revelation. And so it seems that to be deemed "life", our little bag of chemicals must tick all the following boxes indicating that it can somehow absorb nutrients to generate the energy it needs to sustain its own life, respond in some way to stimuli, adapt over time to its environment, and most importantly (and here's the key thing) it must be able to divide and reproduce itself (interestingly, biology is the only field of study where multiplication and division are deemed to actually be the same thing).

And so, our newly emergent little throbbing bag of tricks must it seems in some way be able to pass on genetic material to create another similar living entity, an heir, the next generation, so that life can continue beyond the brief moment in time our little "emergent life" was probably likely to live. Without this our little miracle would be nothing more than a "one-generation" wonder, and Earth's experiment in generating and supporting life would quickly be over before it had got out of the starting blocks.

So let's assume then that we are rejecting the "Mary Shelley method" for creation of life, and that even though the right conditions likely existed for such an event, life was not in fact sparked into being solely by the application of a fortuitous single massive lightning bolt suddenly animating a conveniently located clump of chemicals that just happened to be lying around minding their own business. In which case let's attempt a basic (actually very basic) understanding on what and how life actually comes to be "life". Well, we will need to start with what we are told are commonly known in scientific circles as "the building blocks of life", organic compounds known in science labs as amino acids.

In 1953 a simple experiment was suggested by the Nobel laureate Harold Urey and conducted by one of his students, Stanley Miller, whereby Miller donned some suitably thick rubber gloves, some protective eye-goggles, and proceeded to put a gaseous mixture of methane, ammonia, hydrogen and water vapor into a sealed flask, stood well back, and subjected it to an electrical charge for a week. The idea being to simulate the effects of the constant jolting caused by electrical storms on the equivalent of Earth's early atmosphere and oceans.

The experiment was clearly basic and likely more than just a little dangerous (which is probably why our very smart Nobel laureate had his assistant perform the experiment for him) and was based on massive assumptions around just what constituted an approximation to early Earth's conditions, but nonetheless, after a week or so of such electric shock therapy the flasks had formed a brownish scum, which when analyzed were found to contain amongst other things, several amino acids. It appeared that Miller and Urey had spontaneously "created" organic compounds from a simple gas mixture and an energy source similar to that presumed present on Earth four billion years ago. Understandably the press of the day (never ones given to understatement) was beside itself with reports of how life itself had now been created in a test tube.

Clearly it is not as simple as all that. Demonstrating that life's building blocks could very well have spontaneously emerged out of the conditions that existed on early Earth is an important step in our story, but, creating such primitive amino acids is one thing, however they are not in themselves life. No, their importance in our chain of events that led to the generation of our first simple Achaean life is that amino acids are what we need to create clever little things called proteins.

Without proteins, life could not exist, they are involved in every aspect of every living thing. It's proteins that are required for the structure, function, and regulation of the body's tissues and organs, and are almost limitless in the huge range of roles they perform. Proteins protect the body from harm, they form enzymes which carry out almost all of the thousands of chemical reactions that takes place in our bodies, they are the messengers that transmit signals to coordinate biological processes between different cells and tissues, they help us move and absorb nutrients. They are found in all the tissue that joins muscles to bone to allow movement, and forms the skin that protects us, without which we would all just be a mass of bones, organs, and gunk, lumped together in a random heap on the floor unable to connect or hold ourselves together, we'd be not so much a human body, rather just a very complex puddle. They are the microscopic worker-ants that make us all what we are. In short, if anyone, or anything, is going to create any sort of life, proteins need to be top of the list of ingredients.

And here's where our newly created amino acids come in (fresh from Harold Urey's test-tube), because to create these little protein wonders it seems you need to string specific amino acids together, but (and here's the tricky bit) you need to do so in a very particular order. Think of proteins as a long freight train, and amino acids as the individual carriages hooked up together to make that train. Now there are millions of different types of proteins, the human body itself needs hundreds of thousands of varying types of proteins to ensure our continued daily existence, and each type of protein has its own very unique sequence, of a very specific number, of a very specific set of amino acids. Clearly then, making all these much needed very unique proteins out of our newly created amino acids was not going to be an easy "throw it all together with some sticky-tape and glue" type project.

So it seems that if life was going to emerge, step one was going to be rustling up some of these rather clever little proteins, but given that we must assume that there was no one hanging around at the time with a convenient set of flasks, pipettes, Bunsen burners, and a protein-cookbook, whose sole job was to sit around and rustle-up such complex proteins all day, are we really then expected to believe that all these newly created amino acids all just miraculously, and spontaneously, assembled themselves into the correct chain of specific amino acids in the correct order, all in very fortuitous, but very random events? Clearly the chance of this happening is less than that of a virgin birth or peace in the Middle East, so how do we explain the creation of all these little wonders?

Well the answer appears as disappointingly unscientific as it is simple. What the science geeks tell us is that the answer is simply that proteins just want to be. Amino acids do not appear to combine (or to use the correct geek-speak, "polymerize") randomly, rather it seems they are to a large degree self-ordering and naturally form highly specific sequences, specific sequences that just happen to make our much needed simple proteins. Such self-ordering should not happen, there is no available explanation for it, but we are reliably informed by the same science geeks that somehow it just does. This is of course all very convenient for our journey towards creating life, one of our key components it seems just snub their noses at both science and logic and just want to be, like death and taxes it seems, they just happen. And with this rather unscientific turn of events it seems that planet Earth was now a giant step closer to life, it now had some simple proteins, and with that it seems the chef had his first important ingredient.

However, it would be too much to assume that all proteins just spontaneously appeared all at once, all readily available for the creation of all life-forms (that would be too much to ask even for our miraculous self-ordering amino acids). No, what our clever scientific community of bearded bods believe is that complex proteins (those with more than just a handful of amino acids in their little "protein train") evolved much later as simpler proteins gradually joined (polymerized) over the course of time into the more and more complex protein structures we see today. However, simple life-forms require only simple proteins, and that first spontaneous emergent life was a simple as life could get, and thus for now at least the pots were on and simmering nicely.

But let's not get too carried away, we're still not quite there yet, even with this miracle of "self-ordering" amino acids into simple and evolving proteins that help perform many of the functions that are needed to support life, we still do not have any actual life. Remember we still need a way for our potential emergent life to somehow reproduce itself. As amazing as proteins are, this is not a trick they have yet mastered. No, for this it seems we need a magical little chemical called deoxyribonucleic acid, or in terms you and I can more readily recognize, DNA. DNA is the hereditary material that can perform the magical trick of making copies of itself. Actually it is all it does, and as such it is a replicating black-belt ninja master.

However, it wasn't until the early 1950's that anyone was even aware of DNA and its "ninja" secrets, and all thanks to the pioneering work of 4 very unlikely scientists, Maurice Wilkins, Francis Crick, Rosalind Franklin, and James Watson, who apparently for the most part struggled to work together as a team or even manage to like each other enough to remain on speaking terms, but who somehow contrived to collectively unravel the nature of DNA and it's now familiar double helix shape, and the part it plays in the complex process of cell replication.

In fact DNA and the whole process it masterminds remains something of an overly complex "black-box" to those of us with IQ scores that can be counted on just fingers and toes, but seemingly it's a process that involves the intricate partnership between DNA, as a "recipe book" for genetic codes, our good old proteins to help action the replication, and a second funky little nucleic acid known as ribonucleic acid (RNA) which acts as the translator and messenger for the whole process such that the worker proteins can understand the genetic code that the DNA is expecting it to duly help replicate.

But given that this process is indeed complex enough such that science bods have only really been able to understand it's nature since the middle of the last century, it was clearly something that was going to be far too advanced in evolutionary terms to have evolved, fully formed and ready to go right out of the box at the time our first emergent life was considering what bag of goodies it needed to actually spring itself into life. So, if our newly emergent life was not to be afforded the luxury of today's complex DNA process to allow it to replicate, how then was it expected to perform this little, but somewhat vital trick?

Well, the solution it seems was discovered in the 1980's when it was found that our earlier mentioned little messenger friend, RNA, turns out itself to actually have the capability to store very simple genetic codes and to catalyze proteins to action their replication. RNA could actually do the job all on its own (admittedly only in a very simplistic "it's not pretty but it will do" kind of way) and this led to the surprising hypothesis that perhaps just a simple form of RNA was in fact the initial genetic system of life's early organisms. This hypothesis places RNA front and center when life originated, where self-replicating RNA molecules acted as a sort of simplistic Jack-of-all-trades for cell replication, and as such led to this early period of life's evolution to be not unreasonably known in certain scientific circles as the "RNA world".

Now RNA left on its own can get the job done, however its use as life's "replication tool" was going to prove something of a problem later down the line, you see RNA is extremely fragile and makes the genetic storage and replication process of anything other than the very simplest forms of life prone to mutation and error. Thus, while evolution kept its life-forms very simple (actually very simple) RNA could do the job, but as soon as evolution started to get a bit creative it's clear that RNA was soon going to be found a little light in the toolbox. So it seems that ultimately, RNA was gradually relegated to focus its attention solely on the role of messenger as the far more sophisticated process of replication using DNA evolved with its far greater stability, and therefore its ability to build far longer, less fragile, genomes (genetic codes), and thus enabling the range and complexity of a single replicating organism to be greatly expanded.

Effectively, within the early "RNA world" you got the local odd-job man to come over and slap some paint on your ceiling, with DNA you got Michelangelo to turn up and paint a ceiling that resembles the Sistine Chapel. But at this point in our story it would appear that the crude use of our odd-job man, RNA, is all our simple emergent life had available 3.8 million years ago, and indeed it seems all it needed to enable it to perform the rather necessary task of replicating itself.

So we now have our required proteins to help our emergent life fidget around a bit and absorb nutrients, and we have a basic tool for replication, all we need now is a convenient little package to wrap everything into, and we have something that if we are very lucky, will spark itself into life. That package is called a "cell", and it is cells that are the foundation of life. Only cells have ever "lived", it is only within the safety and nurturing confines of a cell that amino acids, proteins, DNA/RNA, and all their intricate components, actually come together to constitute life. Outside of a cell, everything else is about as alive as my old brown shoes or this book you are holding.

The cell is the basic biological unit of all known living organisms, and is the smallest unit of life that is classified as a living thing. The cell was first seen by a certain Robert Hooke in 1665, who with the use of the coarse microscopes of the day, identified little chambers in plants that he went on to name "cells" because apparently they reminded him of the cells monks of the day would live in (clearly, 17th century monks must have been an austere bunch and not too big on home comforts).

Subsequent improvements in microscopes led to the discovery of microscopic life-forms such as bacteria and protozoa, and such observations led to the development of what biology Gurus call "cell theory". Cell theory is the widely accepted explanation of the relationship between cells and living things, and states that all living things are made up of one or more cells, that new cells are created by old cells dividing into two, and that cells are indeed the fundamental unit of structure and function in all living organisms. The theory holds true for all living things, no matter how big or small, from bacteria to Blue whales, from simple amoeba to Albert Einstein. If you are deemed to be "alive" by the great powers that be, at the basic biological level then, you are merely a vast collection of fortuitously cooperating cells.

And so some 3.8 billion years ago out of some rather smelly primordial sludge all the necessary ingredients for the spontaneous generation of life all fortuitously came together in the nurturing and protective warmth of a cell, the switch was flipped, and life miraculously emerged. It was the simplest form of life you could imagine, it was only a single cell, but it was a tiny chemical factory that absorbed nutrients, could replicate itself, move and evolve, it was a self-contained unit of life.

Such simple single-celled organisms and are known in classroom textbooks as "prokaryotes", as a living organism they aren't very much at all, they did very little, they exist only as a single cell, they are minute, but they were life, and most importantly of all, they were now alive and (figuratively at least) kicking on planet Earth.

4

The Common Ancestor Of Us All

And so, some 3.8 billion years ago, the universe had created a rather convenient set of conditions on our newly formed planet, sent in its top chef who did some clever things with amino and ribonucleic acids, set the gas on high, stirred the pot occasionally, and miraculously a very simple single-celled life-form emerged. His job done, our chef duly departed (presumably to go and repeat the whole process on some other unsuspecting planet somewhere out there in the vast Universe) leaving his latest creation behind to fend for itself. Life on planet Earth had emerged, but what happened to it next was now going to be controlled purely by the twin forces of evolution's random crap-shoot and pure luck. From this point forward, effectively anything (or presumably absolutely nothing) could have happened.

However, clearly not wanting to rush things, initially evolution did very little at all with planet Earth's only living inhabitants, and so for roughly the next 2 billion years or so these simple single-celled organisms (a simple Prokaryotic cell) remained Earth's only life-forms, feeding on toxic sludge, and seemingly happy living the single-celled simple life. For these prokaryotes life was good, no hassles, plenty of food, an ever-growing population, and not much need to do anything other than just relax, kick back, and enjoy life. If there were couches around at the time they would all have been happily sat on one and not moved for the next 2 billion years.

However, sometime within that first two billion years of their existence, one particularly progressive and forward thinking prokaryote did a rather clever thing, and it was something that would not only change the face of life on Earth, but also change the face of the planet itself. It was also, as we shall see, going to have a rather unfortunate effect on most of its prokaryote cousins.

It seems that one day our little game-changing cell decided that feeding on the toxic chemicals in the primordial soup was no longer for him, and so instead turned to a new, more satisfying source of energy. The source of this new menu item was a heady cocktail of sunlight and the hydrogen which was freely found in water. It had somehow found a way to absorb water molecules, and extract the energy it needed from sunlight via the use of the hydrogen the water contained, and then (and here's the game changer) it released a new and soon to be a rather important gas (one we now all collectively call "oxygen") as the waste by-product of the whole process. In his search for better menu items, our little prokaryotic friend had inadvertently invented *photosynthesis*, and it was to be the single most important evolutionary step in the history of life on Earth outside of its first emergence.

These early bacteria-like prokaryotes that now obtained their energy through photosynthesis are known as blue-green algae (or more correctly cyanobacteria) and they turned out to be a huge evolutionary success, primarily due to the fact that their use of sunlight as an energy source had created a much more efficient way of making energy. They were very good at what they did, and evolution generally rewards such high performers, and so cyanobacteria were soon an abundant life-form on Earth.

As these cyanobacteria proliferated so naturally did their production of waste, which rather fortuitously for us all was being released in the form of oxygen, which was thus now conveniently finding its way into the atmosphere. At first almost all of this "new gas" was absorbed by the Earth's rocks, primarily iron, effectively inducing a "mass rusting" of the Earth, and as the smart people out there who know their rocks (the Fred Flintstones of the scientific world) will tell you, this led to the huge deposits of iron oxide which formed the banded-iron rock formations that provide much of the world's iron ore today.

Eventually though, the cyanobacteria became so prolific at pooping out oxygen that even the ever-welcoming iron became so saturated with all the gaseous waste pouring into the atmosphere they could no longer capture all that was being produced, and so at this point, about 2.4 billion years ago, with nowhere else to go, oxygen started to accumulate in the atmosphere, and it had an almost immediate effect. So much so that this point in earth's history, the point where biologically induced free oxygen started to accumulate in the atmosphere, was deemed important enough to also be given its own name, and so has been known ever since as the "The Great Oxygenation Event".

However, this "great" event is also known in some more "glass half empty" circles as "The Great Oxygen Catastrophe" for reasons that may not be initially obvious. But as you will remember, outside of our new and friendly cyanobacteria, all other life up to this point had been happily living in a world without oxygen, and obtaining their energy from a steady three square meals a day of primordial soup, effectively still living off the Earth's old menu. For such anaerobic (non-oxygen using) organisms, oxygen it appears turns out to be rather toxic, and so as oxygen levels slowly continued to grow in the atmosphere (courtesy of our all new super-efficient cyanobacteria) it became increasingly more lethal to these innocent little bystanders, and thus over time our "Great Oxygenation Event" successfully started to wipe out most of Earth's anaerobic inhabitants in one foul swoop.

This has now afforded our little cyanobacteria life a unique place in earth's history as having the honor of being responsible for both the invention of photosynthesis and the oxygenation of our atmosphere, but also the rather dubious additional honor of being solely responsible for one of the most significant extinction events in Earth's history. Quite the achievement for something that is about 100 times smaller than a pin-prick.

As these oxygen-pooping cyanobacteria continued to do their thing, and the oxygen levels in the atmosphere started to increase, another new and rather exciting thing also started to happen. Just as the existing anaerobic prokaryotic cells were being unceremoniously consigned to the recycling bin, the planet was being introduced to a whole new type of cell, one that was far more sophisticated that the simple prokaryotes, one that had miraculously found a way to harness all this new found oxygen in the atmosphere, and one which would ultimately lead to you and me.

These new kids on the block are formally known to biology students and scholars alike as Eukaryotes, and these little miracles of evolution were a cellular version of a Rolls Royce when compared to their more basic, "horse-and-cart" prokaryote cousins. The main distinguishing feature of these new cells was the introduction within the cell itself of distinct "compartments", each of which are enclosed inside their own protective membrane, within which full attention could be focused on performing specific metabolic activities for the cell. These little self-contained "toolboxes" within the cell are known as *organelles*, and the two most important of these were the nucleus (a specific compartment to house and protect the cell's fragile RNA), and mitochondria (a fancy science name for the cells new state-of-the-art power generators which in a huge evolutionary leap were now fully converted over to Earth's new energy-efficient gas, oxygen).

It is this compartmentalization with its protecting membrane that allowed the more complex, yet far more reliable and stable process for cell reproduction to evolve and now be carried out by the specialized use of DNA as the genetic blueprint. As such, along with the newly evolved mitochondria, eukaryotes were now able to greatly increase their efficiency and stability, and thus increased the potential for cell diversity, a diversity that could not have existed under the old rather unstable and unreliable simple "horse and cart" prokaryotic anaerobic-based cells.

But just how did these new "Rolls Royce" cells evolve? Well it is the widely accepted belief of those we trust to know this kind of stuff that eukaryotes and their "compartmentalized" organelles emerged as a result of *endosymbiosis* (for those of us who struggle to walk and chew gum at the same time, this is just a clever way of say "joining together") of existing individual prokaryotic cells. Endosymbiosis theory states that several key organelles of eukaryotes were formed as unwitting free-living simple prokaryotes merged, invaded, captured, were absorbed, or just simply decided to co-habitat, and found that living together suited them both much better. As one cell was taken inside another, each new co-habitant eventually learnt to specialize in a certain function of the cell, and did so now from the safety of its own little compartment within their resulting new, merged, co-habitating home.

The bearded gurus who have guided our hypothesis so far suggest that over time these co-habiting single celled organisms became so dependent upon each other they could no longer survive independently, which eventually led to their merging into one, far more complex, single-celled organism. Hence was born the mighty eukaryotes.

These hugely sophisticated organisms were a huge leap forward for life on Earth, and once established, something they did very quickly as oxygen levels in the atmosphere continued to increase, they set the scene for life's next evolutionary miracle. It was to take a little time, about another billion years or so, but eventually these single-celled eukaryotes themselves learned to join together to form multi-celled organisms, and once mastered, the flood gates opened for complex multi-cellular entities like us to become possible.

To form such a multi-cellular organism, cells needed to identify and attach to another cell. As more and more cells joined together, they learned to work together for their mutual benefit, and eventually particular cells evolved to specialize in only certain chemical processes needed for the organisms survival, and in doing so were thus able to create even more specialized, more efficient, and more varied functions for larger and more complex organisms. All species of animals, plants, fish, trees, just about every living thing that can be seen with the naked eye, are an organism made of a multitude of specialized cells all working together to create a unique living organism.

Look in the mirror and you quite reasonably just see yourself, we may not always be pretty, or the best physical example of our species, but whatever you see reflected back at you, you will see as a single whole living organism. However, at a microscopic level you will find you are actually just a collection of about 100 trillion cells, divided into about 200 unique different types (if we were to take every human being alive today, at last count approximately seven billion of us, and each represented just one human cell, collectively we would still only represent about the number of cells to be found in just one of your big toes). Every single one of those 100 trillion cells has a very specific job to do to ensure you remain a healthy, moving, eating, seeing, hearing, breathing, pooping entity.

It is the continuing cell specialization and division of labor among the individual cells within a single multi-cellular organism, that have led to the huge complexity and diversity that makes up our present day multitude of living species. Put trillions of these little cellular miracles together, and somehow all the trillions and trillions of resulting random chemical reactions miraculously all come together on a grand scale, all working together to ensure the seamless operation of all the necessary functions that remarkably adds up to the fully functioning complex life-forms we see today, trees, whales, flowers, you and me.

So now, roughly 3.5 billion years on from the formation of our planet, the single-celled organisms that had been to that point the only form of living organisms to grace its surface, had now learnt the amazing trick of joining themselves together to cooperate collectively, and thus opening the door to an endless possibility of multi-cellular life-forms. This incredible little feat provided just the shot in the arm evolution needed, and from that point on multi-cellular life simply exploded. It was like evolution had just taken a huge hit of cocaine.

First out of the box were sponges, closely followed by cnidarians (jelly fish-like things to you and me), which were the first organisms to display an actual body of any definite form and shape. Fish quickly followed, some of which promptly developed teeth and jaws, and then some (who had apparently become bored with life in the ocean) applied for shore-leave. Once granted, they quickly developed a backbone to support their weight outside of the natural buoyancy provided by water, a means to breathe air directly rather than having to filter water for oxygen, and finally something a little more rigid to walk around on other than fins. Having duly accomplished this triple evolutionary step, those interested in a more land-based lifestyle were renamed amphibians, cashed in their shore-leave, and duly walked ashore about 370 million years ago.

It was then just a series of short evolutionary hops via some interesting side-projects such as the dinosaurs, a handful of mass extinction events, and an ice-age or two, such that around seven million years ago Earth witnessed the start of an evolutionary chain for a particular species (at the time a small, hairy, tree-dwelling creature with an appetite for nuts and berries) which would ultimately lead to you and me.

Within the space of just under 4 billion years, that first emergent life, that tiny single-celled bag of chemicals that had spontaneously emerged out of a very fortuitous set of circumstances on the infant planet Earth, had now evolved into the multitude of living species that we see today. It may well have been sacrificed along the way once evolution decided that oxygen was the preferred way to go, and in all honesty as a living entity it was about as useful as a chocolate teapot, but it represented the miracle of the spontaneous generation of life on planet Earth, and maybe most importantly of all, it represents the single common ancestor of us all.

5

No Stone Left Unturned

It is truly amazing to think that over the course of around 3.8 billion years, life on Earth has somehow managed to evolve from one solitary single-celled, bacteria-like, twitching little bundle of chemicals that was effectively just one step up from nothing at all, into the mind-blowing diversity of animals, plants, and microscopic creatures that we see today. It has been both a miracle and, as we have seen, something of a seemingly inevitable by-product of the conditions that existed on Earth all those years ago. Either way, through a series of evolutionary twists and turns, extinction events, and some downright lucky breaks, we have ended up with a diversity of life on planet Earth rich enough to have kept the likes of Richard Attenborough in full-time employment for more years than he probably cares to remember informing us about it all.

It is a diversity made even more miraculous when we consider it all exists within a very narrow zone enveloping the Earth in just a 13 mile band. The Earth itself is approximately 8,000 miles thick from its centre to the surface (about the distance from New York to Cape Town), but drill a hole just a mile down from that surface and you won't find a single living thing. Look up much beyond the height of Mt. Everest and all you'll see if you are lucky is a bird or two flying somewhat out of their comfort zone, and probably lost.

Quite amazingly, given all the diversity of life we see on our planet, every single living species, from the highest flying bird to the deepest living worm, are all to be found only within that remarkably narrow 13 mile band, It is the only zone within which our planet is able to support life, and thus it is the zone within which all life out of necessity must exist. This zone is known as the "biosphere", and it stretches from the depths of the deepest part of the oceans (about 7 miles down) to about 6 miles above sea level.

Compared to the enormous extremes of temperature we see in the universe as a whole, life as we know it can only survive in a very limited range. Temperatures in the universe can vary by millions of degrees from the depths of empty space to the center of stars like our sun, from an Eskimo's handshake to a Turkish wrestler's jock-strap. However, for life to emerge, evolve and be supported, it seems it can only do so in the presence of water, and thus can only exist in that narrow temperature range that supports water as a liquid, roughly in the range of zero to one hundred degrees centigrade. Outside of this range most metabolic processes that involve water in its liquid form will either melt, freeze, explode, implode, or in some way put its hands up in the air wave a white flag and just give up, thus rendering life, at least life as we know it, impossible.

Water is the vital medium supporting all the characteristics of living things, growth, nutrition, reproduction, it supports the complex network of internal chemical processes unique to life, and it is a key component of all cells. This means then that life is restricted to the upper limits on land of around the permanent snow-line, (slightly higher for those that have wings), and to a depth of the ocean floor. Between its highest and lowest limits, the biosphere on our own rocky planet is thus only around 13 miles thick, a surprisingly tiny range but one that supports a quite extraordinary number of diverse living creatures.

However, just as extraordinary is the fact that it seems we actually have almost no idea of the actual number of different living things we share our planet with. Ask any bod in the street to estimate how many living species there are in the world today and your likely to get an answer somewhere in the tens of thousands. Scientific protocol means that we must appropriately reduce any estimates that include reference to the likes of such factually dubious creatures as unicorns, leprechauns, the Loch Ness monster, fairies, goblins, and the Yeti, but nonetheless, you will still end up with a number that is very likely several zeroes short of reality.

The surprising truth (for most at least) is that estimates from scientific number-crunchers even for just planet Earth's current eukaryotic species, seem to range from about three million to a staggering one hundred million different species in existence today, although it seems those stirred to passion about such things believe the true number is likely around ten million. Either way, it seems the reality is we have no idea of the real numbers outside of the proverbial finger in the air educated guesses.

Even more extraordinary is that fact that in the 250 years since we started to care about such things, we have currently only documented around 1.2 million species living today. That means that even on a conservative scale, there is still likely over eighty percent of land species and over ninety percent of marine species out there yet to be discovered.

So why should we care about such thing? Well, apart from the fact that it's always a good idea to know just what creatures you are sharing your proposed real-estate with, knowing how many plants and animals there are on our planet helps us to understand just how much diversity we can lose and still maintain the eco-systems that we, humanity, depend upon. Given our current track record on such things as greenhouse gases, de-forestation, and overpopulation, this type of information may well prove useful to have tucked away in our back pocket if we wish to survive much past the current millennium.

Unfortunately however, based upon the numbers above, it seems that the bods amongst us tasked with identifying and documenting our planet's species, collectively known as Taxonomists, are unlikely to provide clarity any time soon. At the current rate of species identification and documentation, it will be around another 500 years or so before the job is even close to being done. Now given that each of Earth's species must have a sustainable population, and that it appears we have only identified something around ten percent of the total number of species, you'd think it wouldn't be too arduous a task to just go out find the remaining ninety percent of species we seem to have somehow overlooked. But it clearly is not as easy as you would first suspect.

The problem is two-fold. Firstly it's not like the world is overflowing with trained taxonomists who once a month will don their safari suits, nip down to their garden shed and dust off their butterfly nets, notebooks, and magnifying glasses, and set off into the depths of the jungle for a couple of weeks of new species hunting. Taxonomy is not as glamorous as one would imagine, it's far less the world of Indiana Jones and far more a world where years seemingly need to be spent at a desk carefully analyzing, measuring, and documenting just one single tiny creature to ensure it truly is a new species and not just a strange looking, mutant variation, of an already documented one. Taxonomist's are by necessity a patient breed.

Secondly, the world is a very large place, and most of its inhabitants are very, very, small and so easily missed. It's unlikely that there are any largish animals left to be accounted for (particularly those based on land) and given our human population of almost 7 billion spread across almost every square inch of the planet it's hard to believe that anything big and hairy out there would not have already bumped into one of us somewhere. So to discover the remaining species we would likely need to meticulously crawl around the jungles of the world on our hands and knees armed with handy-dandy magnifying glasses, turn over every rock, shake every tree canopy, trawl every ocean, and sift through the dirt and sand in every distant corner of the planet, to find them. Not an easy task, even for the ever patient taxonomist.

It's also worth more than a fleeting nod to the fact that, as we have seen, life on earth has been evolving for over 3.8 billion years and during this time an untold number of species have graced this world with their presence, and most have rather less gracefully long since been moved to the debit side of evolution's ledger book. From studies of fossils, some 100,000 species of extinct animals have been identified by men with magnifying glasses, chisels, and tiny hammers, but this is clearly just the tip of a very large iceberg. It is estimated that over the course of the 3.8 billion years that life has existed on our planet, around two to five billion different species may well have existed. We can never know the true number, but it does provide some context in that, if true, it means that well over 99.9 percent of all species that have ever lived on our fair planet are now consigned to the history books, and based on these numbers (even for the most optimistic amongst us) the odds don't look particularly favourable for our own little species.

6

Evolution Not Quite Explained

Despite billions of years of evolution, it seems that even today nature still manages to throw up examples of species who display such a level of stupidity that it would imply that their days amongst our planet's living may well be numbered, and thus by implication, that evolution itself is still very much a work in progress, and a process that sometimes still gets it all just terribly wrong.

Take for example the tree climbing kangaroos of New Guinea, a species which on the face of it seems a reasonable evolutionary idea, but on closer inspection may be an idea that should never really have gotten off the drawing board. Look at your average Australian kangaroo, scale it down, put it in a tree, and watch what happens. These admittedly very cute little animals simply do not have the physiology required for running around in trees, and as a result are forever simply falling out of the very trees that for some inexplicable reason they are stubbornly determined to live in, all with the obvious consequences that befalls any object hitting the ground from a great height.

Or how about the African Wildebeest that once a year gather together in one giant stampede and migrate south in search of food, itself not an unreasonable act. But their migration inevitably requires the crossing of several crocodile infested rivers, into which the hunger driven Wildebeest all continue to blindly throw themselves in an effort to cross to the other side, despite seeing their compatriots ahead of them slowly being picked off one by one by the waiting crocodiles who themselves frankly just can't believe their luck.

And probably in nature's most extreme example of collective stupidity (although the jury is still out as to whether this is indeed fact or just urban myth) we have the Lemming, which for no apparent reason will occasionally collect together in large numbers, and then seemingly on the count of three all just run head-first off the nearest cliff plunging to their deaths in what seems to be an inexplicable mass suicide pact. Evolution clearly has something of a less than 100 percent success rate, yet on the other hand we should of course applaud its willingness to push the evolutionary boundaries and try some "outside of the box" experiments.

Even with our own supposed millions of years of assumed progress and evolution, as humans, we ourselves are apparently still capable of producing individual samples of our species that defy belief in their ineptitude, lack of common sense, and downright stupidity. They are shining examples as to why even our own little gene pool could still do with a little added chlorine of its own.

There are many documented cases where such individuals have kindly contributed to human evolution by self-selecting themselves out of the human gene pool by prematurely shuffling off their mortal coil as a result of such stupidity, and in doing so not only confirm their overall genetic unsuitability for inclusion in our select genetic pool in the first place, but at the same time also greatly improving our species' chances of long-term survival.

Consider for example the homeowner attempting to demolish a large brick garden shed who succeeded in his primary objective, but suffered collateral and terminal damage when the cement slab roof fell on top of him as he had unwittingly chosen to stand inside the structure to carry out the said demolition. Or the man who tied helium balloons to his lawn chair to prove he could fly above his neighbor's yards, only to realize to late that his plan to gently return to earth, which involved him shooting enough balloons with an air rifle to control his descent, would unfortunately only result in a two foot cratered impression of himself in said neighbor's yard. And then there is the case of the man who decided to go sliding down an

Italian ski slope one night, riding on padding that he had removed from the safety barriers at the bottom of the run. Unfortunately, it did not occur to him until the final seconds before his untimely death that it might be just a tad dangerous to sled down the exact same slope from which he had stolen the protective padding.

And thus the gene pool is now in far better shape with the removal of these individuals, and we can safely add to the list anyone who attempts their own home electrical repairs, chainsaw jugglers, people who go over waterfalls in a barrel, take their toast out with a fork, poke sticks a Grizzly bears, try to teach themselves to fly, or think they can apply make-up while at the same time driving a car.

Now such individuals help us clearly demonstrate that evolution and the theory of "natural selection" is clearly alive and well, and still actively thinning out the human herd. But actually what is evolution, what do we really mean by natural selection, and why even after millions of years does the herd still need thinning?

Several times in our recent past men with far greater intellect than most have tried to explain evolution, mostly revolving around the idea of organizing evolution as a kind of ladder of lower to higher organisms. The Greek philosopher Aristotle was the first to take a stab at a meaningful explanation when he defined his "great chain of being" which outlined an evolutionary ladder from minerals to plants to animals to man to demons to angels to God. Aristotle suggested that the "form" of an organism, its defining characteristics, was transmitted through the male semen (which Aristotle also saw as merely purified blood) and the mother's menstrual blood, which together interacted in the womb to direct an organism's growth, in a kind of form defining stew.

The *Charaka Samhita*, an 4th century text on traditional Indian medicine, states that the characteristics of a child were determined by four factors, firstly the mother's reproductive "material" (whatever that may be), those from the father's sperm, those from the diet of the pregnant mother (presumably a healthy diet led to an evolutionary "healthy" individual, while a diet of pies, chips, and chocolate, would supposedly bring forth an evolutionary dud), and finally the karma of the particular soul that enters into the newborn (Indian culture saw mankind as a dual being, having a physical body which merely provided an earthly home for the eternal soul).

But for well over a thousand years the commonly accepted explanation across much of the western world for the abundance of life that we see on our planet was that God (he of the large grey beard, long white gown and sling-back sandals) had created all of Earth's creatures, placed them exactly where he wanted them and put man firmly at the top of the earthly evolutionary ladder where he could be served by all the other "less intelligent" creatures. All creatures were seemingly created by some divine act of "spontaneous creation", and so remarkably also just seemingly appearing one day out of the blue as fully evolved species, so it had always been, and so it would always be, and thus was the explanation for life on Earth (Amen). This remained pretty much the belief until the 1700's when a French naturalist with the impressive title of Georges-Louis Leclerc, Comte de Buffon (1707-1788), was the first to argue that life actually had something of a history.

According to Buffon, life originated as a set of distinct types that he called "internal moulds" that organized the organic particles that made up any individual creature, effectively a set of "ingredients" with an accompanying recipe book that was used to concoct all of Earth's creatures. Buffon went on to explain that as a species moved to new habitats, the local supply of organic particles (the ingredients) that was to be found on the shelves of the local "Species R Us" superstore was likely to change, and thus out of necessity the particles would thereby change a species' mould a little (effectively a little tinkering with the recipe). Buffon was, in other words, proposing a sort of evolution based upon adaptation to the prevailing environment.

For example, using only the "local ingredients" available, the recipe for a bear may well result in a Grizzly version in North

America but a Panda type in China, although heaven knows what local mix-up in the recipe may have caused the hippopotamus, the warthog, fish with swords where a nose should be, or a shark with a hammer stuck on the end of its head.

Buffon's theories were visionary yet doomed, primarily because they were based on the relatively skimpy evidence that eighteenth-century naturalists had at their disposal. Yet his theories foreshadowed some of the most important developments in natural sciences in the decades that followed his death. It may be true that not one single idea of Buffon's has withstood the test of time, but his work was still a milestone in science because he thought about the Earth and life in ways that few had before, he was the first to propose that both life and the Earth indeed had a history of their own.

Then in 1801, another French naturalist with the even more impressive title of Jean Baptiste Pierre Antoine de Monet, Chevalier de Lamarck (1744-1829), took a great conceptual leap and proposed a full-blown theory of evolution. Lamarck (as he was simply known to his friends in the interests of time and convenience) was struck by the similarities between many of the animals he studied, and it led him to believe that life was not fixed, and that as environments changed so to organisms had to also change their behaviour in order to survive. Lamarck surmised that if organisms began to use a particular organ more than they had in the past, it would naturally increase in size and complexity in its lifetime.

For example, if an elephant stretched its nose to reach for water in a lake or river enough times (presumably so it avoided the risk of falling in and drowning), what Lamarck called a "nervous fluid" would flow into its "nose" and over time make it longer. Its offspring would then inherit this longer nose, and such continued "evolutionary" stretching would make it longer still over the course of several generations, (presumably to the point where the nose grew so long it now needed its own name, a trunk). Lamarck also surmised that as organisms adapted to their surroundings, nature also drove them inexorably upward from simple forms to increasingly complex ones. Like Buffon, Lamarck believed that life had begun through spontaneous generation, he claimed that new primitive life forms sprang up throughout the history of life, but were continually "evolving" up the ladder of evolution from simple forms to more advanced and complex organisms.

Lamarck was thus proposing that life took on its current form through natural processes, not through any divine miraculous interventions, but a process that had a purpose, one where all species seek to evolve to a higher more complex state. Lamarck's thinking not unsurprisingly became known as Lamarckism, and is really an hypothesis that states that physical changes in species acquired over the life an individual (such as a bigger nose developed through repeated stretching) would actually be transmitted to the individual's offspring, a theory that today is known by the more familiar term, "inheritance of acquired characteristics".

Two key ideas were thus incorporated in Lamarck's theory. Firstly the idea of use and disuse, being the idea that body parts used more often in an organism's lifetime become stronger or larger (such as noses eventually needing to be renamed as trunks), while parts not used will slowly waste away and ultimately disappear (such as a chicken's wings, now only of any use as a menu item at a BBQ). The second was the theory of inheritance of acquired characteristics, the idea that any changes acquired in an organism's lifetime would also be passed on to its offspring, as in the case of the elephant and its soon to be renamed hooter.

Not surprisingly he was mocked and attacked by many other naturalists of his day who were disturbed by the theological implications of his work, believing as they did that nature was a reflection of God's single-handed design (and thus nothing to do with stretching noses or under-used appendages for flight). Shunned by the scientific community, Jean Baptiste Pierre Antoine de Monet, Chevalier de Lamarck died in 1829 in poverty and obscurity (but presumably still with an impressively wide tombstone). But the notion of evolution did not die with him, and as we will see, the idea of evolution was soon to be changed forever. To school children and scientists alike, the now accepted explanation for the evolution of life is inextricably tied to one man, Charles Robert Darwin.

Darwin once said that his life would make a great novel. Actually I just made that up, but he probably would have done so had he known all the twists and turns his life would take before it was over. His is a story of adventure on the high sea, secret theories, and friendship, all set against a backdrop of revolutionary times, a kind of *"Gone with the Wind"* for scientists. But ultimately it is the story of one man's own battle over 30 years against his own demons, to bring the world a theory that explained everything about life, its complexity, its beauty, diversity and elegance.

So if you're sitting comfortably …..

7

The Survival of the Fittest

Charles Robert Darwin was born on February 12, 1809, in Shrewsbury, England. He was the second son of his overbearing doctor father, Robert Waring Darwin, who sent the 16 year old Charles to study medicine at Edinburgh University in the vain hope of curtailing what he considered his son's "idle ways" (19th Century English for lazy teenager), while at the same time fuelling a dream that he would eventually follow in his own footsteps as a Physician.

However Darwin Senior's plans soon hit a roadblock as Darwin Junior quickly discovered he had little stomach for either the sight of blood or the practices of a pre-chloroform surgical world, neither being ideal for a budding 19th Century doctor. What he did discover however was a keen interest in the natural world, and the ever shrewd Darwin Senior realizing that a "Plan B" was quickly needed for his weak-stomached son, switched him to Christ's College Cambridge to study for a life as a rural cleric, believing that the church was likely a better calling for an aimless wannabe-naturalist such as young Charles. At this time the Anglican Church dominated every part of life in Victorian Britain including the scientific world, as such the leading naturalists in Darwin's day were all also clergymen of the Church of England, so his father's move seemed a shrewd one at the time.

Having dutifully completed his theological studies in January 1831, Darwin returned home to Shrewsbury to find a letter waiting for him from his Cambridge mentor J. S. Henslow. It contained an invitation to sail on a trip around the world serving as a naturalist on the HMS *Beagle*. As a young botanist keen for adventure and travel, this was a dream come true, and actually given the alternative

Darwin faced (a future serving as a country Clergyman with only gout and the chance to study butterflies and bugs in his spare time to look forward to) clearly he was going to jump at the chance of adventure on the high seas. Charles Darwin's life was about to change forever, and so it turned out was that of the future scientific world.

The Beagle sailed from England on December 27th 1831, but once underway Darwin was to quickly discover a rather unwelcome bonus of the trip. It appeared that he was more than just a little prone to sea-sickness, an ailment that was to unfortunately stay with him throughout the Beagle's five year journey. However, despite the constant revisiting of the previous night's supper, the journey itself was to prove completely enlightening for Darwin, with each stop The Beagle made on its globe-trotting journey providing him with new insights into the natural world and the forces that drove it.

Strangely, as part of his understandably limited luggage (life aboard The Beagle was a little cramped to say the least) Darwin had with him a copy of Sir Charles Lyell's presumably riveting read, *Principles of Geology,* which set out concepts of land slowly rising or falling over immense periods. Lyell (1797-1875) was the leading Botanist of his day and who's view on geological change (known as "Uniformitarianism", a name which is almost as long as the processes it looked to explain) was an idea in stark contrast to the commonly held belief of the time that geological change was only driven by abrupt changes, known as "Catastrophism", and which had been conveniently adapted to chiefly support Christian beliefs such as Noah's flood. Lyell's interpretation of geological change as the steady accumulation of minute changes over enormous spans of time, became a powerful influence on the young Darwin, and the idea was to form a cornerstone of his later theories.

In September 1835, The Beagle made land (likely much to Darwin's relief) at the Galápagos Islands, where Darwin found a natural treasure trove of plant and animal life, and where he collected numerous weird and wonderful plants and animals that were new to him, including finches and other "strange-beaked" birds which to Darwin's keen eye seemed much alike yet somehow differing slightly from island to island.

Sailing further on to New Zealand and Australia Darwin witnessed how the Norwegian rat, which had been inadvertently introduced by European settlers, was destroying the indigenous "down-under" rats, and how human diseases brought by the same settlers, smallpox, tuberculosis, influenza, measles, whooping cough, (another rather unwelcome gift-basket from the visiting guests) were all taking a toll on the indigenous Aboriginal people who to that point only really needed an immune system that could deal with snake bites and sunburn. Thus Darwin started to witness what he saw as the ongoing and ever-changing brutal struggle for existence and competition for survival that all animals (humans included) had to endure.

After nearly five years at sea, The Beagle finally arrived back in England on the 2nd of October 1836. The voyage had filled Darwin's head with the wonders of nature, and more importantly his experiences and observations had planted all the seeds that would later become his theory for the evolution of all life on Earth, a theory which would ultimately revolutionize science, but a theory that would also almost destroy him.

However, on arrival back at home port the merry crew of The Beagle returned to find a country experiencing great change and progress. Victoria was about to take the throne and the British Empire was growing. Driven by the Industrial revolution workers were flooding to the cities for jobs in the new factories and struggling to live in over-crowded and squalid conditions, while radical ideas from revolutionary French-types across the sea were also starting to find voice in the new British working class who were now openly ready to embrace new ideas such as evolution as a tool for political and social reform.

Thus, against this backdrop of industrial and revolutionary change, Darwin quickly became very aware that his newly forming ideas around evolutionary forces were likely going to be perceived as dangerous, not so much in the intellectual sense, but dangerous socially and politically because evolutionary theory suggested that everything can change, and in the eyes of the establishment this would put him front and center as a radical revolutionary who preached anti-establishment, and more dangerously, anti-Church ideas. Not really a role a plant-loving naturalist who was just happy to be back on solid ground and away from the scourge of sea-sickness was looking for.

It's important to understand that in Victorian England, Science, politics, and religion were all mixed up together, such that to be properly scientific was to be properly Christian (scientists of the day were more likely to be found wearing a white dog-collar rather than a white lab-coat) and any thoughts of evolution outside the doctrine of God's creation of life was damned as anti-Christian, and thus anti-Science. Even just speculating about evolution in these times was clearly going to be a very bad career move for anyone if done so in public. If Darwin was going to pursue his ideas around evolution he would clearly need to do so very carefully.

One of the first things that Darwin did on his return was to seek the help of the renowned ornithologist John Gould (1804-1881) to examine the finches and "strange-beaked" birds he had collected from the Galapagos Islands, and several months later Gould reported back to Darwin with what would be surprising and significant findings. Gould reported that these birds were actually each a unique species of finch, all strictly confined to the Galapagos islands, but also all mutually exclusive from island to island. It appeared that each species had evolved in a different way on each island. This news made Darwin very excited (all in a very restrained Victorian gentleman kind of way of course) as he saw this as significant proof that one species could indeed evolve into another given the right conditions, and enough time. But the big question that remained was how, how did such evolution of species occur?

The first clue was to come in Oct 1838 when Darwin chanced upon a book by the political economist Thomas Malthus (1776-1834) entitled *An essay on the principle of population*, in which Malthus outlines his theory that human population when unchecked would go on and double in size every 25 years or so, a situation that leads to the population quickly exceeding food supply (something that is never going to be a good thing for a growing and hungry population), and leading to what was to be named as a "Malthusian catastrophe".

Such population explosions would naturally lead to a desperate struggle for survival and in which Darwin now sees a mechanism to drive evolutionary change. He saw that as supported by Malthus' theory, a particular species will always breed beyond available resources, but that favorable variations in a particular individual would clearly make it better able to survive and thus also be able to pass on the variation to their offspring, while organisms with unfavorable variations would likely not survive and thus not pass on the presumably "dud modification". Darwin would call this process "natural selection".

By 1843 Darwin had managed to create a 35 page rough outline of his "secret" theory. However quite incredibly, still not daring to go public with his thoughts, his revolutionary ideas were to continue to consume Darwin for another 15 years before he was reluctantly ready to share his masterwork with the unsuspecting world. Darwin was clearly not a man taken to do things by impulse.

In 1844, as Darwin continued with his long and thorough formulation of his theory, (mountains could likely form and erode in less time than it would take Darwin to be completely happy with his work, if ever), Darwin was hit with a sizable bolt out of left field when an anonymous author published their own book on evolution named *"The Vestiges of the Natural History of Creation"*.

The book suggested that everything currently in existence had developed from earlier forms, the solar system, the Earth, rocks, plants, fish, mammals, and ultimately mankind, in short everything had first existed as, well, something else. It explained the initial origins of life by spontaneous generation, and then via a series of progressive steps from simple to more complex organisms ultimately leading to man, with the Caucasian European unashamedly as the pinnacle of the process. The implication of the book (apart from it obvious racial under-current) was that the evolution, leading progressively to humanity, was part of a preordained plan that had been woven into the laws that governed the Universe.

But where Darwin was obsessively cautious to prove and provide the scientific evidence for his theory, the anonymous author of *"Vestiges"* boldly dived straight into print caring little about the small matter of providing any such supporting evidence. As such, the book was unscientific and full of factual errors, but it caught the imagination of the public and became an instant best seller.

The book stunned Darwin, he understood that as a piece of scientific work it was based on little fact or substance (as a scientific work it was in fact about as meaningful as a mouse fart in the midst of a hurricane), but it inadvertently had made public what he had already been secretly theorizing for almost 10 years, his evolutionary cat was now firmly out of the bag and running around causing chaos with the public and church alike. To make matters worse the scientific establishment (the Church) naturally railed against the book as anti-God and immoral, and thus absolutely underscored for Darwin the dangers of what he was doing. Darwin was now like a deer caught in headlights and had no idea what to do next.

In the end he turned to the one person whose judgment he trusted completely. Joseph Dalton Hooker (1817-1911) was a botanist who like Darwin had voyaged to the southern hemisphere (and also likely suffered the same severe sea-sickness). Darwin had approached Hooker on his return to England to help classify the plants that he had collected from his voyage, and from that point on the two struck up a life-long friendship. Hooker was the only person that Darwin had dared share his initial thoughts on natural selection with, famously remarking to Hooker that telling him was like "confessing a murder". And so it was Hooker who figuratively kicked Darwin up his Victorian backside and back into his laboratory again after the shock of "*Vestiges*".

In 1851, as Darwin continued to toil over his great theory, England was celebrating the opening of the Great Exhibition in London, a showcase for imperial Britain's technological might, and the country was in the mood to celebrate science. The man who most embodied this era of change was a certain Thomas Henry Huxley. Huxley (1825-1895) was a charismatic, witty, clever, yet very confrontational man, who like Darwin and Hooker, had journeyed to the southern seas. Huxley was a radical whose primary goal it seems was to challenge and ultimately overturn the existing scientific establishment, and Darwin it seems was just the sort of person Huxley wanted to associate with (the difference being that while Huxley wore his radicalism on his sleeve, Darwin was keeping his firmly under lock and key in the closet). The two eventually met and instantly hit it off, and subsequently forged a friendship that was to have a significant impact on both their lives.

By 1853 Darwin had turned his attention to a crucial aspect of his theory, how to account for the prolific distribution and variation of animals and plants around the world. Orthodox science of the time (and actually for the past thousand years or so) stated that it was God who had created and then placed each species in one particular place and that once there species certainly never had the temerity or the ability to change into new a species. As we have seen, Darwin believed otherwise, but he needed to find a mechanism to help support his theory. Darwin believed plants and animals could move across the world by wholly natural means (any species with either legs, wings, or fins, can travel without the need for any celestial game-board placement) and in a constant struggle to survive in their new found homes they eventually often evolved into new species.

To support his theory Darwin set about proving that seeds could survive in salt water for extended periods of time to demonstrate how they could be transported by ocean currents to new locations, however, as Darwin was studiously beavering away watching seeds in a bucket of salt water for days on end, on the other side of the world an obscure character named Alfred Russel Wallace (1823-1913) had also been looking at the world around him and pondering many of the same thoughts that had occupied Darwin's keen brain for the last 20 years.

In 1854 Wallace had left England for South East Asia, but this was no academically sponsored expedition, his venture was pure business, Mr. Wallace was an entrepreneur. He supported himself by collecting exotic animals and plants from the far reaches of the globe and then selling them to scientific institutions and wealthy patrons around the world with more money than sense, all at exorbitant prices (a kind of Donald Trump meets Allan Quartermain). But Wallace's world was about to collide with Darwin's not because of his entrepreneurial pursuits, but because Wallace it seems was also very interested in evolution.

In February 1855, while busily collecting suitably exotic plants and animals to sell to the rich and gullible, Wallace wrote a paper entitled *"On the law which has Regulated the Introduction of New species"*, in which he discussed his observations around the geographic and geological distribution of both living and fossil species, something that was to become known as "Biogeography". Although the paper makes no mention of any mechanism for evolution, it strongly hints that such a mechanism exists, and clearly indicates he was starting to think very much along the same lines as Darwin, and more importantly starting to come to similar conclusions.

Darwin read Wallace's paper but strangely didn't seem to make a connection with his own work, and so ignored it. But his friends, Huxley and Hooker, saw Wallace's paper as a threat and urge him to publish his own theory quickly before Wallace himself managed to join up all the evolutionary dots and publishes the theory himself, and they convince Darwin that his time may now have come and he must summon the courage to step out of the shadows and formally put quill to paper, write, and then publish his theory.

However, Darwin first decides to continue to connect his own series of dots, and so spends the next year or so studying of all things pigeons, and in particular how man has changed them enormously over the years by selective breeding. He is looking for a mechanism to provide proof of how selected traits within a particular species can be passed down through numerous generations, eventually leading to the creation of an entirely new species. Darwin wants to try and show that what the dedicated pigeon fanciers of the world have achieved over years of selective breeding, what Darwin calls "Artificial selection", mirrors what nature does in the wild via "Natural selection".

Finally, in May 1856, happy that he now has his evolutionary dots connected and all the pieces of the puzzle firmly identified and in place, Darwin starts on the long and arduous process of writing his book on his theory of evolution, initially titled *"Natural selection"*. But committing the book to paper proves difficult, Darwin's compulsion to be so meticulous, so authoritative, explain every single detail with supporting evidence, means that it starts to look like the book when completed will likely be the size of a small house, but nonetheless, Darwin presumably orders several wagon loads of paper, a few barrels of ink, and duly sets himself to the task.

Then, 2 years into what was turning out to be something of a literary marathon, fate once again steps in with its size 13 boots to provide another painful blow to Darwin's sensitive region. In June 1858, Darwin receives another surprise letter from Alfred Wallace, who is still industriously securing exotic species from the jungles of Indonesia, and included with the letter is a 20 page essay in which Wallace uses natural selection as his core explanation for his theory of evolution. While Wallace's paper, entitled *"On the tendency of Varieties to Depart Indefinitely from the Original Type"*, did not actually use Darwin's term "Natural Selection", it did however outline the mechanics of an evolutionary divergence of species from similar species due to environmental pressures. Wallace it seems had completely independently stumbled upon his own missing piece of the evolutionary puzzle, and now clearly saw his own version of evolution via the process of natural selection.

Darwin was likely choking on his toast and Victorian marmalade as he read Wallace's paper, and instantly realizes that his last 20 years of work, the work he was currently meticulously committing to paper ready for publication as his own theory of evolution, may now all have been for nothing, Wallace may well have beaten him to it. If the first essay received from Wallace three years earlier raised Darwin's eyebrow only slightly, this latest paper ensured that both eyebrows were now firmly raised skyward as high as his Victorian brow would let them.

Darwin turns to his friends, Huxley, and Hooker, and in despair throws the whole matter of Wallace's essay and what to do with it into their hands. They quickly realize that Darwin must publish his theory first, before Wallace, else it will be "Evolution by Natural Selection" by Alfred Russel Wallace, and no one would even know of Darwin's own great theory and near 20 years of meticulous (although admittedly secretive) research.

Eventually they came up with a cunning plan, a plan made even more cunning by the fact that Wallace was blissfully unaware of what was happening sat as he was on the other side of the world. The plan required Darwin to first quickly write an abstract of his work to date (itself no small feat given Darwin's usual glacial speed in putting pen to paper), so that he can publish it alongside Wallace's essay.

Huxley and Hooker know that scientific convention states that the first public reading of even an abstract paper establishes ownership of the theory, so they hastily arrange for a reading of both Darwin's abstract (now duly written and at something approaching warp speed in comparison to Darwin's usual glacial style) and Wallace's essay at the Linnaean Society in London, one of the world's leading societies for the study of biology, and crucially (in yet another cunning move) with Darwin's paper presented first followed by that of Wallace. It was on the back of just such Victorian ingenuity that Britain had forged an entire empire.

But now Darwin is in an inescapable position and faced with a sobering thought. He must now follow-up his hastily published but necessarily brief extract, and publish his theory in full, and do so quickly. So Darwin settled down with the lengthy half-finished manuscript he had already spent over two years composing, and set about squeezing it into a single condensed book that was to become "*On the origins of Species*". At last the theory he had been wrestling with for over 20 years now had a good shot at being written, and in a concise, readable manner that would not require the reader to invest in a forklift truck to carry it home and half a lifetime to read, and in doing so would turn Darwin into one of the most influential scientists of history.

At the heart of the book is the idea that since the Earth evolved life has been in a state of constant change, introducing the idea that not only could species change into other species, but that given enough time even simple bacteria could eventually evolve into rabbits, fish, trees, plants even stoic Victorian scientists with impressively large beards. It goes on to explain that this evolution is achieved via a brutal and unrelenting struggle for survival, and through this cruel struggle for survival evolutionary change occurs through natural selection as individuals within a species out-compete rival individuals within the same species, reproduce, and thus drive evolutionary change. Natural selection it seems favors, and preserves, the traits that help an individual adapt better to their natural environment, and so survive better. Using Lyell's geological theories as his basis, the book goes on to explain that over vast stretches of time these favourable traits add up until descendants are ultimately very different from their earlier ancestors, and in this way they can eventually evolve into altogether new species.

In the fall of 1859, 13 months after starting his great work, an exhausted Darwin finally lays down his quill. Notably, within the book's still lengthy 502 pages, Darwin includes only a single passing line about the one species that was likely to cause more controversy than any other. Darwin realized that as soon as he published people would naturally be talking about the obvious implications to our own species, to human evolution, and so quite deliberately he did not directly speak to human evolution at all in the book until the very end, and even then he only states "Light will be thrown on man and his history".

As such, Darwin was indirectly making it very clear that humans were very much part of his evolutionary view, without actually committing to the obvious links to human evolution head-on, and thus hopefully skillfully side-stepping any controversy that any direct reference to man's own evolution would likely bring. However, his attempted side-step was not to be quite as successful as hoped, and the implications of his theory would prove just too great to slip by unnoticed.

On Nov 24th 1859, *"On the Origin of Species by means of Natural Selection"* finally went on sale in London. Sold for the bargain price of 14 shillings it sells out quicker than life-jackets on the Titanic and is an immediate hit. Sales were clearly bolstered by the curiosity and expectation that had grown around the book's pending release, but also by the fact that although a scientific work and highly scholarly in nature, it did not require a brain the size of a planet to read. It was accessible, full of new ideas, controversial and dangerous in nature, a book for everyone, and one guaranteed to put a smile on the face of any publishing house editor.

Importantly, as a scientific work it seemed to also speak to the people and the times. People were already starting to see that the Church was becoming more and more out of step with its increasingly "modern-thinking" congregation, and that it no longer seemed as relevant to the masses in their new urban life as part of the great Industrial revolution, and Darwin's book was a shining example of that modern thinking. It was the penny and the pound that was driving the good folks of England now, not the pulpit.

The reviews from the scientific (religious) establishment however were, as expected, not quite as enthusiastic, and the book was met by much gnashing of teeth and wailing from the pulpit. Much to Darwin's dismay reviews focused on the one point that he had taken great pains to try and avoid, referring to the book as a theory of "men from monkeys', and so kick-started many a heated debate in the parlours of Victorian England around whether man was indeed a divine creature or merely just another "evolutionary beast" as Darwin's book had "clearly" implied.

Clearly such debate was not going to just conveniently go away, and the long expected public confrontation with the Church based scientific establishment eventually came to a head in June 1860 at a meeting of the British Association for the advancement of Science, at the Oxford University Museum. Running the meeting was Samuel Wilberforce (1805-1873), a geologist and Bishop of Oxford, and son of William Wilberforce the great British slavery reformer. Wilberforce was not a man given to embracing new ideas quickly or easily, and who regarded evolution as nothing short of religious heresy. For Wilberforce it was absolutely essential that he and his fellow Christian scientists had complete separation between man, who was made in God's image (and thus clearly a "special" being), and animals which were produced on Earth merely to serve man and his needs. And in the other corner was Darwin's great friends Huxley and Hooker, who were attending in support of their friend's theory, but also pushing evolution as an example of the future for British scientific thinking.

The debate began with Wilberforce taking the stage and drawing a distinct line between humanity and animals and insisting that "lower" organisms do not, cannot, somehow become intelligent beings, asking (not unreasonably) "does a turnip become a man?". However the real confrontation occurs as Wilberforce attacks Huxley directly by asking him whether he was, according to his evolutionary beliefs, related to the ape on his Grandfather's side, his Grandmother's side, or both? A potential knockout blow.

But Huxley's corner had clearly done their homework on his opponent, and he responds firing on all cylinders, noting to the gallery that Darwin's book does not directly raise the matter of man's evolution from ape to human at all, and that it was thus patently obvious that Wilberforce had not even read the book, at fact that if actually true was a bit of a schoolboy error from Wilberforce. Huxley quickly followed up this counter-punch by landing a swift one-two combination by stating that if he had to make a choice he would be proud to acknowledge the ape over someone who used his logic to deceive and mislead the public, and who remarked on scientific questions on which he clearly had no real knowledge. Wilberforce was put on the canvas and quickly counted out.

Huxley and the young reformists had put down the religious scientific establishment and made them look red-faced and foolish. It was a huge victory for Darwin and his supporters, his once heretical ideas formed over a 20 year period had now found a level of acceptance, and in doing so heralded the start of the end for the old church-driven scientific order.

Forgotten in all the excitement of publication and debate of course was that all these hugely significant events were all happening without the express knowledge or involvement of one of the theories' own co-discoverers, Alfred Wallace, who was still seemingly buried deep in a far flung corner of the jungles of Indonesia. It would be another 3 years before Wallace finally returned to England, where he now found a country that had undergone huge cultural and scientific change since he had left all those years ago. Changes that he had in part unknowingly helped set in motion with Darwin.

On his return, Wallace would finally get to meet Darwin, but surprisingly rather than claim his rightful position as co-discoverer of "Natural Selection", revel in the glory of his role in such a scientific gift to the world, and wallow in the adulation of his peers (something the that the ever gracious Darwin was more than happy to support), he incredibly positions himself instead as a supporter and strong advocate for what he openly calls "Darwin's great theory". Maybe having spent so long isolated in the jungles of Asia Wallace hadn't quite grasped the significance of his own work, or maybe he was just happy taking a seat in the dug-out, but either way it is Darwin's name that fills the pages of today's high-school text books on evolution, not the Darwin and Wallace double act.

Charles Robert Darwin died on the 19th of April 1882. Thanks to the petitioning of his friend Huxley, he was buried in Westminster Abbey close to Isaac Newton, a very suitable resting place for one of the great men of his time. In the course of his lifetime he had successfully changed the way human beings thought about themselves, providing nothing short of the explanation for the existence of every living creature. All in all then, not a bad life's work for someone who at the age of 21 was staring down the barrel of a life as a country clergyman, preaching to the poor and chasing butterflies in his spare time.

8

The Answer is in the Peas

Darwin had successfully outlined a theory that explained how evolutionary forces over the course of millions of years had taken us from that single common ancestor to you being here today wondering what on earth you were doing still reading this book. But what Darwin was unable to explain was the mechanism by which his "descent with modification" actually worked. How do "modified" traits actually get passed down from generation to generation? If one of your parents was an Olympic sprinter, chances are you will be out-running everyone in the school yard too, Darwin had explained why, but what was still left unanswered was how.

This passing of traits through subsequent generations is known today as heredity, or genetic theory, and it explains why you look like your father (and if you don't it explains why you should, and why you probably also need an urgent conversation with your mother), but for early Victorian scientists how heredity actually worked remained as puzzling as an early Bob Dylan lyric or an IKEA instruction booklet, and a particular source of frustration to Darwin and many other overly bearded boffins of the time. After all, heredity lies at the heart of Darwin's evolutionary theory, the variations in each generation are the raw material for natural selection, Darwin had the right theory, but was missing the actual mechanics.

Common scientific belief at the time was that individual traits were passed down through generations as species mated, with the traits from each parent somehow "blending" together in their offspring. However, this idea of blending inheritance created a problem for Darwin as any "beneficial" variance in traits passed from one parent would be rapidly lost as it became progressively more "diluted" by the constant blending that would occur with each new generation. Blending as a process for heredity effectively made evolution by natural selection implausible, so either the notion of blending traits was wrong, or the followers of Mr. Darwin needed to go back to the drawing board, and until this little mystery was solved the mechanics of Darwin's theory remained a frustrating mystery.

Ironically, just as Darwin was publishing the *Origin of Species*, far off in a secluded monastery in what is now the Czech Republic, a monk named Gregor Mendel (1822-1884) was studying hereditary, specifically hereditary in of all things garden peas (don't ask). Now just why a monk would want to study garden peas in the first place we may never know, you would have to think it was not something so pressing that Mendel's God felt it necessary to hand down such a task (presumably either via some mystical vision or moment of enlightenment) to one of his dedicated followers above say feeding the poor or healing the sick, so it was clearly a calling Mendel had chosen for himself, and fortunately for those scientists who were still scratching their heads about how heredity might work, he did.

In fact, it appears Mendel was so taken with peas that he selected over twenty different varieties of them, bred thousands upon thousands of pea plants, systematically inter-breeding them, and recorded how certain traits were passed on from one generation to the next. Mendel observed and recorded several traits that today's green-fingered wellington-booted horticulturists will gleefully tell you are easily recognized, and ones that rather conveniently can only occur in one of two forms in the common garden pea, this included the color of the flower (being only either purple or white), the seed color (being only yellow or green), and the seed texture (being only either smooth or wrinkled).

Through this selective cross-breeding of common pea plants over many generations, Mendel discovered that surprisingly traits showed up in offspring without any blending of parent characteristics at all, the colour for instance was always either wholly purple or white, their seeds either wholly yellow or green, and that in all cases no "blended" textures (slightly wrinkled or not quite as smooth) appeared in the offspring of cross-pollinated peas at all, effectively putting pay to the "blending of traits" theory, at least in peas anyway.

In fact, what Mendel observed was that when he first cross-bred smooth and wrinkled peas, he produced only peas that were all smooth. As a result Mendel radically proposed that each of the peas produced had actually inherited both the smooth trait from its smooth parent and the wrinkled trait from the other, but only the smooth trait became visible. The smooth trait it appeared was "dominant" over the "recessive" wrinkled trait. Mendel was clearly on to something here with his breeding peas.

When Mendel then produced a further generation of peas from these new offspring, it appears that a quarter of the peas were now found to be wrinkled. Mendel surmised that in this next generation, with each parent pea now holding one dominant (smooth) and one recessive (wrinkled) trait, the traits were again similarly handed down, but this time around a quarter of the new peas inherited two recessive (wrinkled) traits from its parents which made them of course wrinkled. It seemed each new pea was randomly inheriting from each parent just one of either the dominant (smooth) or recessive (wrinkled) trait. Statistically therefore (and I'm assuming you can do the math here), one in four of the new peas was inheriting a recessive trait from both parents, and thus without inheriting the dominant smooth trait at all, those peas had no choice other than to become wrinkled.

Now this being the late 19th century and Mendel being tucked away in some far flung corner of Europe in a secluded monastery on some remote Czech mountain, of course he had no notion at that time of genes or chromosomes (such revelations would come many years later), but he had stumbled on the key that would unlock the door to such study, and in today's world of genetic science Mendel's results are now summarised into what's known as the *Law of Segregation* and the *Law of Independent Assortment*.

In modern terms, and in a language that even those of us whose genetic heredity did not include the gene for "boffin", the *Law of Segregation* states that each individual has two copies of each trait, (what we now know as a "gene" or group of genes), that produces a certain characteristic. These two copies are now called alleles, and the offspring of two presumably willing parents receive only one allele from each. *The Law of Independent Assortment* states that there is no mixing between the genes for different characteristics, they remain separate entities, with one of the two received alleles always being dominant over the other. As offspring, you will always have exactly the same traits as the dominant alleles received from one of your parents. In short, if your mother has a large pointed nose, and your father a flat snub nose, you should not pin your hopes on a career in modelling.

Mendel presented his remarkable and ground-breaking results in 1865 in a paper entitled *Experiments in Plant Hybridisation* (but obviously in German), and absolutely no one paid any attention to it whatsoever (maybe because it was in German, or maybe because no one was really interested in peas). Disillusioned, and now with several thousand peas he didn't know what to do with, Mendel abandoned his experiments and turned his attentions to more monk-like activities such as feeding endless supplies of pea soup to the local poor. He died in 1884 having never met or exchanged notes with Charles Darwin or any of his peers, and was remembered only as a puttering old monk who had a liking for breeding peas and plants, and making pea soup. It was only some 15 years after his death that scientists realized that Mendel had in fact revealed the answer to one of life's greatest mysteries.

However, as interest in genetics grew at the turn of the 20th century many of these first new "geneticists" finally rediscovered Mendel's ground-breaking work. But these new "Mendelists", as they were now catchily called amongst the more hip of the scientific community, still faced a problem with reconciling Mendel's theories with that of Darwin's theories of natural selection. Darwin had talked about natural selection gradually altering species by working on tiny variations over long periods of time, but Mendel had outlined major differences between traits passed to new generations (a pea being either smooth or wrinkled) and nothing "gradual" in between, and thus rather inconveniently somewhat at odds with Mr. Darwin's great theory.

However, by the 1920's several scientists such as J.B.S Haldane, Sewall Wright and Robert Fisher, were doing clever things with a slide-rule and a chalk-board building mathematical models that were able to show how natural selection could indeed operate in a "Mendelian" world. Their approach became known as "Population genetics" and their research helped explain that any given trait was in most cases the product of many genes working together rather than just one single gene (as in the very simplistic case of Mendel's peas), and a mutation to any one of the single genes involved would likely only create very small changes to the trait rather than some drastic transformation (peas suddenly going from wrinkled to smooth, or your rather ugly, overweight neighbours who are susceptible to skin complaints suddenly creating offspring who are all swimsuit model material). They also revealed how mutations arise and, if they are favored by natural selection, they can then spread throughout a population.

These population genetics boffin-types with their lab coats and eye glasses held together with sticky tape outlined how the evolutionary process is actually driven by four basic mechanisms for evolutionary change. Each of the four mechanisms being capable of altering the frequencies of genes in a population, and as a result they are all capable of driving descent with modification (evolution).

First off their chalk-board was the idea of *gene mutation,* a random change in the DNA sequence of a gene. DNA is the hereditary material of life and affects how an organism looks, develops, behaves, its color, its physiology, how big its nose will be, and everything in between. So any slight change in a gene's DNA sequence can cause changes in any aspect of its life. Mutations can be beneficial, neutral, or harmful to the organism, but mutations do not "try" to supply what the organism "needs". In this respect mutations are completely random, regardless of whatever particular mutation may occur, it is completely unrelated to how useful that mutation may, or may not, be.

The only mutations that matter to the evolutionary process are however those that can be passed on to an organism's offspring via reproduction, those that are hereditable. But remember mutations can only occur in the DNA of genes that already exist within the organism, neither mutations, wishful thinking, nor a lifetime of prayer, will gives pigs wings or turn turtles into mutant ninjas.

Next to jump from our genetic boffins notebooks was the mechanism known today as *gene flow*, also known as *gene migration*, and is the movement of genes between sub-populations of a species. In nature a species is often divided into multiple local sub-populations, with the individuals within each sub-population freely mating at random (among willing participants of course), but naturally less frequently with individuals from other sub-populations due to geographic distance, ecological barriers, or even cultural or physical preferences. Thus, when individuals from different sub-populations mate less frequently, gene flow of the overall population is restricted, and may lead to certain sub-populations becoming genetically quite different over time. However, when genes are carried to a population where those genes are far less common or previously did not exist, gene flow can be a very important source of genetic variation.

Consider the Vikings of Scandinavia, who rather rudely took it upon themselves to invade Britain in the late 8th century. They landed on the beaches of the northeast coast of England and over the course of the next couple of hundred years conquered most of Britain. However, along with their horned-helmets, longboats, and axes, the Vikings also brought with them their blue eyes and blonde hair, and thus also became the primary reason why there is now far more blonde hair and blue-eyes amongst the children of modern day

England than there would have been had the Vikings decided to stay at home all those years ago.

The mechanism of *Genetic drift* concerns changes in gene frequency that are driven merely by random chance and Lady Luck. As an example, consider a population of birds happily living on a small island in the Pacific Ocean, let's call these birds Lesser Spotted Drifters. Now let's assume that roughly half their population has five white spots on their tail and the other half only three. Now the population usually all got on together very well, until one day a little misunderstanding over worming rights resulted in a large number of the population with only three spots on their tails deciding they needed a little time away from the others, and so they flew over to a neighboring island for a little break. While there, out of the blue, a large meteorite fell on their island getaway unfortunately wiping out the entire island's inhabitants, including the vacationing Drifters.

Now the result of this unfortunate random event was that the original population back on their island home now had significantly fewer Lesser Spotted Drifters with three spots on their tails, and thus as a consequence future generations of their species would now very likely evolve with a far greater number with five spots on their tail, maybe even over time eventually removing the three spotted variation from their gene pool all together.

These random chance changes in a population's gene pool is known as genetic drift, and happen to all species, but is most important in small populations where a loss of genetic diversity is even more likely given the small number of the population with which to maintain such diversity. There's no avoiding the vagaries of Lady Luck for any of us, just ask the dinosaurs whose gene pool was more than just a little affected by one such moon-sized random event.

The forth, and last, mechanism for evolutionary change had already been outlined some seventy odd years earlier by our friend and hero Charles Darwin. *Natural selection* as we have discovered is the means by which beneficial variations in a population tend to be preserved while unfavorable variations tend to be lost. A key concept of the theory being that there is variation within populations, and as a result of that variation some individuals are better suited (fitter) to their environment while others are not. Because members of a population must compete for finite resources (food, shelter, and a suitable mate), those better suited to the environment will naturally out-compete those that are less well suited.

Now we should note that this idea of "survival of the fittest" doesn't just mean that if you look like a movie star, are as fit as an Olympic athlete, and have scientific degrees hanging on your wall to testify to your superior intelligence, that you are a natural shoe-in for natural selection. Regardless of being the fastest, strongest, or biggest, the process of natural selection still requires that you produce offspring to pass your genes onto. If your firing blanks in the bedroom for instance, regardless of any other beneficial trait, you will still be of little use to the evolutionary process. But, given ample time, the result of natural selection is a change in gene frequency within a population such that individuals with more favorable traits will become more common in the population and individuals with less favorable traits will become less so (and so likely over time disappear altogether).

However, natural selection is just a process, it is not all-powerful, there is no evolutionary overlord guiding its hand such that species can slowly work towards perfection, neither does it have any goals or in some way strive to produce a balanced ecosystem. It is merely the simple result of variation, differential reproduction and heredity, just like the car you drive or a $10 hooker it is mindless and mechanical.

Similarly species themselves do not "try" to evolve, natural selection cannot try to supply a species with what it "needs". Natural selection just selects among whatever random variations exist in the population, and the result is evolution, not evolution with a purpose. The world's collective genetic boffins do not all come together once a year to debate what would be a nice shiny new adaptation for each of the world's species, publish a paper detailing their recommendations, and then expect nature's evolutionary forces to somehow act upon them.

And so it seems that the genes present in a population of any species are in a constant state of flux due to these four forces, and as such genes mutate, individuals are selected, and populations evolve. It is important to understand that genes and individuals do not evolve, only populations evolve. But an individual's genes mutate, and those mutations often have consequences for these individuals, individuals with different genes are then selected (for or against) and as a result populations change over time, they evolve.

Now if we think back to the notion that we are all descended from the same single-celled common ancestor, we may question whether these four basic mechanisms for evolutionary change could indeed account for the huge variety of life we see today. A process such as mutation might seem way too small to account for the huge divergence and differences between say your cat and the apple tree it sits in at the end of your garden. But actually it's not.

Remember, no matter how small the changes, life has been accumulating such mutations and applying natural selection for nearly four billion years (and that's a very long time even on a galactic scale), and more than enough time to create an evolutionary tree that has continually grown, split and diversified, to produce such a wide variety of species. It appears that Georges-Louis Leclerc was indeed right, life does indeed have a history, and a very long history at that.

Population geneticists had thus now helped prove that evolution is indeed carried out mainly by small mutations over long periods of time, even on the scale that gives us such diverse species as trees, fish, insects and humans, and yes even the hippopotamus, the warthog and hammerhead sharks. This reconciliation between Mendelian genetics, Population genetics, and Darwin's theory of

Natural selection, finally provided a unified explanation for the basic mechanism by which evolution happens.

Charles Darwin had indeed been right all along, and as complicated as the huge diversity of life we see before us seems, there is an underlying simplicity in evolution and its process, a simplicity borne out of the simple truth that every living thing, past, present, or future, is just the result of small increments and modifications going back almost 4 billion years, back to that single-celled common ancestor from which we all began.

Trace mankind's evolutionary history back to that first single ancestor, then trace any other living organism's history back to that same beginning, an apple, a dung beetle, your pet cat or dog, and in every single case our evolutionary paths will at some point cross, a point on the evolutionary tree where we were indeed the same. Understand this and we may all think twice next time go to stamp on that hairy spider in the bathroom, or chop down that tree because it is hindering the view from our bedroom window.

9

A False Start

On December 19th 1912, the world's newspapers led with sensational headlines declaring "Mankind's missing link found", "Darwin's ape found", and "Ape link proven". The headlines were sparked from a meeting the day before of the Geological Society of London, where a certain Charles Dawson, an amateur archaeologist and fossil hunter from Sussex in England, had claimed that a worker at his local gravel pit in Piltdown had some four years earlier given him a skull fragment which had sparked his fossil-hunter's interest. Further revisits to the site seemingly uncovered further skull fragments, at which time Dawson took the fragments to a certain Arthur Smith Woodward, a keeper of the geological department at the British Museum, whose interest also now duly sparked, returned to Piltdown with Dawson between June and September 1912 unearthing yet more skull fragments, some teeth, and half a lower jaw bone.

During that same meeting at the Geological Society, Woodward proudly announced that a reconstruction of the found fossil fragments showed a skull very similar to that of a modern human, but with a brain about two-thirds of the size. The jaw-bone he further explained was almost indistinguishable from that of a chimpanzee, and his findings led him to enthusiastically conclude that this "Piltdown man", or *Eoanthropus dawsoni* ("Dawson's dawn-man") as the specimen was formally named in its finder's honour, clearly represented the evolutionary "missing-link" between apes and humans. With a human-like skull filled with a sizable brain, yet with an ape-like jaw, to Woodward and Dawson it was the only obvious conclusion.

Clearly then, the discovery of the Piltdown man was the evidence needed, the link between man and ape had been found and as it turned out had all along been a resident of Sussex, England. The world's newspapers had their sensational lead story, and family and friends could put the beer on ice as human anthropologists around the world could all now pack-up their bags and come home, Charles Dawson had solved the mystery of mankind's missing-link once and for all.

Now other similarly sensational finds of what were presumed to be fossilized versions of our early ancestors had already been made in the Neander Valley in Germany as early as the 1850's, where bones found in a quarry were presumed to be a type of human ancestor, they were duly named *Neanderthals*, but then spent the next decades being refuted as being less of those of an early human ancestor and merely just the bones of a presumably rather ugly modern human with a somewhat unfortunate brow-ridge and a seemingly remarkably heavy set bone structure. However, despite the claims for them being just "unfortunate" human specimens, what these Neanderthal fossils did provide was some lingering belief that Darwin's missing link, and even the birthplace of mankind, was potentially in an unremarkable corner of Germany.

Thus, it was against this backdrop that British Scientists, desperate to prove that Britain had played its own part in the story of human evolution, saw their evolutionary dreams come true that December day in 1912, with the Piltdown man. Charles Dawson had now seemingly provided the evidence they needed to prove that Britain was in fact the birthplace of mankind, not some remote corner of Germany, and as such the find and its associated evolutionary conclusions were readily accepted with only a cursory nod to any real critical review of their validity. Piltdown man ticked all the right boxes for the eager rather nationalistic British scientists of the day, and so for the next 40 years Piltdown man reigned as the official holder of the title "the missing link", and thus remained the key member of mankind's family tree.

Some 10 years or so later, and half a world away in South Africa, Raymond Dart, the head of anatomy at the University of Witwatersrand in Johannesburg, received a small fossilized skull of what appeared to be a child. The skull had an intact face and lower jaw and remarkably also a fossilized cast of the brain. The skull had been found earlier that year by quarrymen working in a limestone quarry in Taung, South Africa, where it had subsequently sat on the mantle over the fireplace of the quarry's Director until a visiting friend of the family who just happened to be an anatomist and student of Dart, recognized its potential significance, negotiated its liberation from its mantle-place home, and promptly sent it to their eager mentor.

Dart examined the skull and with some subsequent chalk-board wizardry estimated it to be around 2 million year old, noted that the fossilized brain was only slightly larger than that of a fully grown modern-day chimpanzee, and that the skull would have been positioned directly above the spine showing that the creature likely stood upright, a distinctly human trait. To Dart this was all clear evidence of an early ape-like creature evolving distinctly human characteristics, and promptly declared it a new species, naming it *Australopithiecus africanus* ("Southern ape man of Africa"), but the skull itself was soon nicknamed as the "Taung Child".

Now unfortunately, this "Southern ape-man" presented the scientific world, particularly in Britain, with two very large problems. Firstly, the current undisputed "missing link", the Piltdown man, had clearly established that our early ancestors had a large brain and ape-like teeth, while Dart's Taung Child had a brain only somewhat larger than a modern chimpanzee along with distinctly human like teeth, and so rather inconveniently, the exact opposite. And secondly, and even more worryingly for the British scientific establishment, any formal recognition of Dart's *Australopithecus africanus* would have implied mankind had its roots not in England, but in Africa, and of course that would just not do at all. The Taung Child was thus quickly and conveniently brushed off as merely a fossilized ape and nothing more, and so spent the next 50 years serving as a novelty paperweight on someone's desk at the University of Witwatersrand.

Then in November 1953 something quite extraordinary happened. New technology that enabled the dating of fossils more accurately had been developed since Piltdown man's first discovery, and palaeontologist Kenneth Oakley along with Oxford professor Joseph Weiner, both of who had always been rather skeptical of Piltdown man's authenticity, had subsequently undertook a re-evaluation of the early Sussex resident based upon this more recent technology.

To their surprise what they now found was that the jaw and teeth were not of the same age as the skull, and that in fact they were not even fossils at all, just old bones. Indeed, on closer examination some of the bones it seems had been boiled and then stained with chemicals to make them match each other and give the appearance of being fossils. To make matters even worse, the so-called fossilized teeth appeared to have been deliberately ground down to give them the required wear pattern. Oakley and Weiner clearly smelt a rat.

Further CSI-like sleuthing work showed that far from being an early ancestor of man, Piltdown man was actually a composite of three separate species. It consisted of a human skull of around medieval age, the jaw bone of an orangutan, and the filed-down teeth of a chimpanzee. It turns out that the Piltdown man was less of an example of our early human ancestors, and more of a manufactured scientific freak, a kind of "orang-panzee-man".

Oakley and Weiner, armed with their new evidence, were now convinced that the Piltdown man was not only a fake, but was actually an elaborate and carefully prepared hoax. And so, on November 20, 1953, in a bulletin of the Natural History Museum, they reported their findings to a suitably shocked and outraged scientific world. The newspaper headlines the following day shared the story with the rest of the world, mockingly declaring that "Fossil Hoax makes Monkeys out of Scientists".

Weiner clearly had Dawson in his sights as the perpetrator of the hoax, but since Dawson had rather inconveniently died back in 1916 tracking down the evidence to actually pin the hoax on him seemed a little pointless, and besides, many already believed that Dawson lacked the scientific knowledge to actually fake the bones that had deceived so many scientists, at least by himself. And so others were also implicated as the possible hoaxer(s), with the most famous name linked to the forgery being that of Sir Arthur Conan Doyle, the creator of the famous fictitious Victorian detective Sherlock Holmes. Conan Doyle apparently had lived near Piltdown, and as a doctor and well known fossil collector he had the knowledge it seemed to put together the hoax. Indeed, in his book *The Lost World* published the same year as Piltdown man was found, he rather cheekily makes reference to bones being easy to fake. Unfortunately the true identity of the hoaxer has never been discovered, and likely we will never know who perpetrated what has been rather glamorously called "the greatest scientific hoax of all time".

Unfortunately then, it seems that the study of human evolution, a science that was at this point less than a 100 years old, had just spent the last 40 of those going down a scientific blind-alley by basing much of its theory around an elaborate hoax. Not the greatest start for those keen to search for the roots of mankind's evolutionary line, and somewhat embarrassing for science as a whole (particularly science emanating out of Britain). Fossil finds previously ignored or dismissed as merely "apes" or "unfortunate looking" modern-day humans, were now re-examined in a new light, including Dart's Taung Child which is now recognized as one of the most treasured and important fossils in the field of anthropology (the study of mankind), something of an upgrade from a novelty paperweight.

Of course the issue was made worse by the fact that palaeontologists (clever bods who make a living studying the evolution of pre-historic life) are not exactly blessed with a huge number of fossil specimens to work with in the first place. Since our early ancestors decided to split from their ape cousins, step down from the trees and take a stab at creating our own evolutionary line about five million years ago, we would have to guess that some several billion or so specimens representing varying stages of that lineage have walked this Earth. But it seems that to-date our whole understanding of our unique evolutionary history has been based on the evidence gleaned from the findings of just a few thousand fossilized individuals (and not even complete specimens, actually, just tiny fragments from each of those few thousand individuals). It's a little like trying to piece together how a car works when all you have is a tire and a spark-plug.

It is one of the major disappointments of today's scientists standing ready with calculators and microscopes in hand eagerly seeking to piece together our evolutionary history that not all animals conveniently turn into fossils once they've served their allotted time amongst the living. Unfortunately when an animal dies (including our own early ancestors), it's an unfortunate twist of fate that their last contribution in this world is often as constituting a free meal, with their carcass usually being scavenged by predators glad of the easy pickings, or they simply just rot away as a result of acidic soil and bacteria which effectively dissolve away over time all evidence that they even existed in the first place.

To have any chance of being preserved for posterity as a shining fossilized example of their species, our early ancestors would have had to have conveniently died in an area where conditions were exactly ripe for fossilization. This usually means under the fortuitous circumstances (if dying can ever be called fortuitous) where their recently deceased body was quickly covered in sediment or volcanic ash before any hungry local sabre-toothed tiger could sniff them out, and typically such occurrences are rare. Unfortunately our predecessors quite sensibly preferred to live in areas of lush jungle or savannah (because that's where the food was), but it was also where the other scavengers, insects, bacteria, and acidic soil were. As a consequence, fossil remains from such areas it seems can be as rare as rocking horse droppings.

Interestingly, as hard as it is to find such fossils, when they do, the fossil hunters of the world will primarily stumble upon of all things specimens not of bones but of teeth, mainly due to the enamel on teeth which helps to preserve them far better than bone. Should you wish to be remembered some several million years beyond just the next few generations your family will likely care to remember you, the best advice it seems is to invest in some industrial strength fluoride toothpaste.

So as a rule, to ensure you go from living being to fossil, rather than to scientifically unhelpful dust, generally means ensuring you shuffle off your mortal coil in an area of high sediment and aridity, or to do so close enough to an active volcano to ensure you are quickly engulfed by fossil-making volcanic ash, however unfortunately neither being a preferred setting for our early ancestors. And of course even if one of our early ancestors did obligingly clock out under such beneficial conditions, the palaeontologists of today would then still then have to find them, a search that seems to depend primarily on luck rather than any real scientific process. Not surprising then that we have found so very few fossil remains of our ancestors, and thus have had to piece together an evolutionary line based more on assumptions and interpretations than hard evidence.

And then of course to make matters just a little worse the fossil evidence we do have does not all fall neatly into evenly spread examples across our evolutionary line so as to clearly map out our path from ape to human. No, rather unhelpfully they appear randomly, and while modern science can now accurately date when the original living owners of the fossilized bones likely strolled around slowly progressing that evolutionary path, they often offer nothing more than another random and tantalizing piece of a puzzle we are trying to piece together with little or no knowledge of what the final picture is supposed to look like. And just to add a little icing to the top of the cake, in almost all cases the fossils are merely partial skeletons, skull fragments, or random limb bones, a tooth here and a jawbone there, rather than a neatly packaged complete skeleton with a helpful note attached conveniently detailing hereditary line and species. So, all in all, not a lot to work with then if you are trying to piece together nothing less than the history of an entire species.

And so it should be little surprise that each new fossil find rarely fits neatly into any existing accepted understanding of the human family tree. Rather, they offer tantalizing new evidence that often seems at odds with current understanding, and are thus open to numerous and often conflicting interpretations. No wonder that palaeontologists, a proud and fame-hungry group it seems, tend to interpret finds in ways that miraculously supports their own personal theories even if such findings fly in the face of existing published evidence from their peers. The world of human palaeontology if nothing else certainly embraces open interpretation and lively debate, and thus as a result leaves us with a constantly shifting understanding of our human evolution, an understanding of which itself rarely finds general acceptance across the scientific world as a whole. To some extent, the whole scientific field can be viewed as something of a scientific train-wreck, and thus unfortunately a major challenge for anyone trying to piece together just where, and how, our own unique species came to be.

10

The First of Our Line

Despite some lingering contention in the murkier corners of Anthropology HQ, the now current consensus (based upon some clever genetic science, molecular clock dating, and some chalk-board jiggery-pokery) is that the evolutionary line of primates (a class of mammal covering all monkeys, apes, orangutans, lemurs, tarsiers, and of course our own little species) diverged from other mammals somewhere around 65 million years ago. About 5 million years later the orangutans collectively decided they needed their own unique evolutionary line, and then somewhere around 5 to 8 million years ago, first the gorillas decided to also go their own evolutionary way, and then it seems our own ancestral line finally split from our chimp cousins to start our own unique branch in evolution's family tree.

So, it seems that as recently as about 7 million years ago we (humans) were still in the same ancestral line as chimpanzees, up to that point in time we weren't "like" the hairy little primates with a weakness for bananas at all, we "were" them. As such our closest modern-day relatives remain those friendly chimps, with our genetic code seemingly being somewhere between 95 and 99 percent identical (depending it seems on which side of the genetic classroom you choose to stand) to the chimps you will see happily acting out a tea party at your local zoo.

The basic adaptation of our own evolutionary line is not as many might think the ability to moon-walk like Michael Jackson, or even drive a car with one hand while holding a mobile phone or applying make-up with the other, no, the basic evolutionary adaptation for our evolutionary line is *bipedalism* (the ability to walk upright on two legs), with the earliest of our own particular evolutionary line so far having been identified being three species known respectively by the rather catchy names of *Sahelanthropus, Orrorin,* and *Ardipithecus*. All are now thought to be "facultative" bipeds, (which in plain English means that they were bipedal when moving on the ground, but still quadrupedal when moving around in the trees), and it's believed that it was somewhere among these early bipedal species that the particular evolutionary line emerged that would eventually evolve into you and me.

In the early 1990's a research team digging around in the Afar region of Ethiopia stumbled upon the fossilized bones of what appeared to be an early ape-like creature. The various fragments amounted to an extraordinary 45% of the total skeleton, and were dated to be between an even more extraordinary 4.2 and 4.4 million years old. On further examination the fossils appeared to be the remains of a small-brained (about the size of a modern-day chimp or a soccer hooligan), 110 pound female, and included most of the skull and teeth, pelvis, hands, feet and limbs. In all over 100 different pieces of fossilized bone were found (the human body has a total 206 bones according to medical physicians and gangland "muscle", whose job it is to know these things).

This exciting new species was duly named *Ardipithecus ramidus*, from the Afar word *Ardi* which means ground or floor, *pithecus* from the Greek word for monkey, and *ramid* being Afar for root, this catchy but overly grand title thus denoting the belief that the species not only walked on the ground but was also (so they believed) the likely root of the human family line, although the fossil itself quickly became known simply as "Ardi"

Although artistic impressions based on the skeletal remains of what Ardi likely looked like show that she was no great beauty in today's terms, (you wouldn't want her showing up at your front door on the arm of your teenage son), she was of great evolutionary significance, and given her age she clearly could not represent a common ancestor of both chimps and humans, a line which we have already seen is believed to have likely split a few million years earlier. However, although Ardi's pelvis and feet appeared better suited to walking upright, she still appeared to have big toes that appeared opposable indicating she was still likely in part at least quadrupedal. Ardi it seemed still spent as much time romping around in the trees as she did walking upright on the ground.

As we have come to expect from the ever argumentative egos of the palaeontologists and anthropologists of the world, there is still much debate as to whether Ardi and her kind represent the start of the evolutionary line that went on to develop into today's *Homo sapiens*, but regardless, *Ardipithecus ramidus* is still considered an important evolutionary step towards that line, and a such Ardi herself (all 45% of her) remains something of a 4.4 million year old evolutionary celebrity.

Then a few years later a team of French fossil hunters doing French-like things in the Tugen Hills of Kenya (taking 3 hour lunches and cleaning their teeth with garlic) found the fossil remains of what appeared to be another early ape-like creature but surprisingly dating back even further than Ardi, this time to around 6 million years ago. The find was nowhere near as complete as that of Ardi with only some teeth, part of a jaw, part of one arm, a leg bone, and some fingers, however, this was clearly enough for the excited French to declare this new species, (duly named *Orrorin tugenensis* from the Tugen word *Orrorin* meaning "original man" and a nod to the hills where they were found) as a bipedal species and thus at that time the earliest known representative from our own ancestral human line. So far a total of 20 specimens of *Orrorin tugenensis* have been found and all in the same approximate area, mostly by French fossil hunters (and so then presumably mostly found in the late afternoon after a leisurely lunch), and date from between 6.1 and 5.7 million years old.

A year later pieces of a small fossilized cranium, jaw, and some teeth were found in the Djurab Desert of Chad. Once painstakingly pieced together by jigsaw fanatic scientists the find was found to represent a skull (or at least the top half of a skull anyway), which seemed to indicate that the original owner of the fossilized fragments was likely also a bipedal ape-like creature (based on some CSI-like sleuthing showing the location of the opening where the spinal cord presumably met the cranium, and thus indicating that the head must have been held upright). What was extraordinary about this find was that it dated to be approximately 7 million years old.

The skull was duly assigned as a new species and named *Sahelanhropus tchadensis*, after the Sahel region of Chad, *anthropus* meaning "human" and implying its human ancestry, and a dutiful nod to its Chadian homeland, while the skull itself was nicknamed "Toumai" meaning "hope for life" in the local dialect. The find was announced to the scientific community with much fanfare and quickly declared as the most significant find in the field of human anthropology in the last 75 years, all amid claims that it represented the oldest human ancestor after the split of the human line from that of the chimpanzees, and thus was clearly the "founding father" of our human evolutionary line.

However, Toumai not unsurprisingly has been the centre of much controversy, with many claiming that it may be an overly strained interpretation (even for the usual broad licence applied by the discoverers of such finds) to place *Sahelanthropus tchadensis* at the root of our human family tree, based merely upon what is effectively half a skull. Indeed, improvements in genetic analysis and molecular dating have now placed Toumai very close to the time of human/chimp divergence, and so Toumai is now more readily accepted as likely representing a common ancestor of both humans and chimpanzees, but as expected, no real consensus has yet been reached by the scientific community. Effectively, the evolutionary jury currently remains out on poor old Toumai.

Clearly *Sahelanthropus, Orrorin,* and *Ardipithecus,* would have been extremely primitive creatures when compared to you and I, they had brains about 1/5 of the size, were a long way off from discovering the benefits of hunting, fire, stone tools, coffee to-go, or getting a decent haircut. They probably communicated at best in grunts, and likely preferred to pass their spare time checking each other for fleas rather than starring at the night sky contemplating the origins of the Universe, but they had started to walk upright, (probably somewhat uncomfortably, and likely for only part of the time), and were doing so several million years ago. We may never know whether these very early cousins of ours are true direct human ancestors, but what is clear is that as of around 7 million years ago, the evolutionary line that would eventually lead to you and me was out of the starting blocks, walking upright, and well under way.

Of course this view could all change in an instant if some more of those easily excitable French happen to stumble upon the fossil of a whole new early *ancêtre* of ours while digging up the onions in their backyard, but then that's the nature of trying to piece together our evolutionary line. It seems we are always just one new fossil discovery away from having everything we understand to that point being turned upside down and thrown out of the proverbial "evolutionary tree".

11

Stepping Down from the Trees

Becoming bipedal is arguably the single most important evolutionary step in mankind's several million year history, but what made our early ancestors, Ardi, Toumai, and their buddies, decide to step down from the trees and make a stab at walking around on the ground on two legs in the first place? Most of the scary animals with big teeth and matching appetites were roaming around down there, and up there in the trees you would have to presume Ardi and his crew were relatively safe (apart from the obvious fall-hazard). Down on the open ground they would be exposed and so be much more vulnerable, and judging by the fossil evidence found of the type of carnivore that would have found Ardi a pleasing menu item, just about every such animal around at that time was stronger, faster, and better equipped with deadly teeth and claws than our admittedly scary and very hairy looking, but ultimately defenceless, ancestor.

Going bipedal also meant being forced to adopt some radical physical changes too. Over time, these early "stand on your own two feet" pioneers would have needed to evolve increasingly more robust knee and ankle joints, and an S-shaped spine to better support the increased weight, the big toes needed to be moved into alignment with the other toes to help with forward movement rather than climbing (else your feet would rather inconveniently try to take you in different directions), and the top of the spinal column moved to a more central position under the skull in response to their new upright position. However, the most significant and problematic changes occurred in the pelvic region, particularly for the female of the species. The skeletal changes in the pelvis to support an upright stance necessitated a smaller birth canal, and as any of today's modern moms will tell you, this means a significantly more painful birth for the mother, along with an increased risk of complications for both the baby and the mother.

To be fair, evolution responded to this particular gynecological challenge by ensuring that human offspring are born when the head is still small enough (just) to get through such a narrow space, although today's birthing mom's may tell you the process could still do with a little tweaking. But evolving such a solution meant that Mr. and Mrs. Ardi's newborns now arrived physically immature, vulnerable and still helpless, this of course lead to the long-term dependency of offspring on their mothers, which itself required the development of, and a commitment to, a solid male-female monogamous relationships, and as a result the family unit was born (the noble professions of marriage counselling and divorce lawyers can also likely trace their roots back to this time).

All of this makes the decision to step down from the trees and reinvent themselves in a new upright-model seem extremely problematic and risky, and quite frankly on the face of it more trouble than it was worth, so why on earth would Ardi and his fellow tree-dwellers decide on such a course of action? It's a bit of a head-scratcher for sure.

Well as it turns out, the good folks at the anthropology "brains trust" believe that the answer may well be that our early ancestors may not have "chosen" to step down from the trees and reinvent themselves in a new upright model at all. Studies appear to show that around the time Ardi was nervously stepping down from her arboreal home, the Earth was going through a period of major climate change, one that would have ultimately seen the tree-filled jungle home that Ardi knew, slowly turning to savannah. Stepping down from the trees and going fully bipedal was clearly a risky business, but the evidence appears to show that our ancestors likely came down from the trees out of necessity rather than choice. They didn't leave the trees, the trees left them.

However, once down from the trees, adopting a bipedal upright stance did surprisingly also seem to offer several advantages over the "knuckle-walking" option that we see in use in today's modern apes and French rugby players. Firstly it freed up the hands for reaching and carrying food, carrying and protecting the young and newborn, and ultimately leading to the opportunity to fashion stone tools. Being upright is also more energy efficient than "knuckle-walking" and thus enables the potential for running faster and for longer distances (handy if you are being chased by a 500lb sabre-toothed tiger), it also gives a better field of vision, both essential if your plan against the scary predators with the big teeth and claws is to either see them first and hide or to try and outrun them. Being upright also significantly reduced the area of the body exposed to direct sunlight, thus protecting against overheating, again a useful adaptation in the hot open savannahs of Africa (unfortunately for Ardi, the big floppy hat and Factor-50 sun-block would not be invented for at least another few million years). So it seems that maybe Ardi was just a little bit smarter than we initially gave her credit for.

The evidence for such vertical ambulation in those very early ancestors however was always based on the study of the (admittedly scarce) available fossil evidence by overly bright science-types, and so was only really "implied" by how such fossil-geeks interpreted the bone structure and mechanics of such evidence. However, in 1976 much more solid and indeed surprising evidence was to be unearthed at Laetoli in Tanzania, by Mary Leakey a noted paleoanthropologist (the fancy title for someone who studies ancient humans as found in fossil evidence) and her team of intrepid fossil hunters.

Mary Leakey and her team had been scratching away for fossil evidence of early humans at a site known as Laetoli in Northern Tanzania for two seasons (it seems that just like sports teams, fossil hunters also have seasons), and so far had only a few fossilized teeth and bone fragments to show for their efforts. Then one day, quite by chance, they noticed something that took them quite by surprise, what they found was not a fossil at all, but what seemed to be the cast of a footprint embedded in the ground. They excitedly excavated the surrounding ground and unearthed what has become one of the most remarkable discoveries in the study of human evolution.

What they had miraculously stumbled upon was a 25 metre stretch of 54 fossilized footprints embedded in the rock that seemed to indicate that they were made by 2 individuals, walking together, and remarkably judging by the age of the rocks, doing so some 3.6 million years ago. The footprints had miraculously been preserved as the 2 upright-walking individuals, (presumably either two prehistoric ramblers, or a couple out for a romantic early evening stroll), unwittingly left their footprints in muddy, wet, volcanic ash which quickly hardened preserving their tracks, which were then presumably covered over again by further ash deposits, and thus fortunately encasing them in a kind of "man-tracker" time capsule for future discovery some 4 million years later by Mary Leakey and Co.

Analysis of the footprints clearly showed where the heel of each individual had struck the ground first as each foot came down, followed by a clear push-off from the toes as they moved forward, remarkably similar to the footprints you and I would leave behind today on a wet sandy beach. These were obviously expert and steady walkers, (whoever or whatever the owners of the footprints were, they were clearly not drunk), and given the age of the rock the footprints were found encased in, and based on the acknowledged species of the fossilized teeth and bones found in the same vicinity earlier in the season, the general consensus was that they had been made around 3.6 million years ago by members of a species that had been recently discovered by one Donald Johanson, and which he had named as *Australopithecus afarensis.*

Donald Johanson was an American anthropologist, who was part of a team working in the Hadar region of Ethiopia's Afar triangle some 2 years earlier in 1974, and who discovered what was to that point the most complete skeleton of one of mankind's early ancestors. What Johanson found was a remarkable specimen totalling around 40% of an early "human-like" skeleton (the actual percentage of the overall skeleton remains hotly disputed amid claims of some "creative" counting around the individual bones found), which was estimated to be around 3.2 million years old. The original owner of the skeleton was deemed to be an adult female, around three and half feet tall, weighed about 65 pounds, apparently would have looked a little like a chimpanzee, and from the fossil evidence seemed to be capable of walking upright.

Johanson announced the find as a new *hominid* species (a fancy scientific term for any modern or extinct bipedal primate), named the species *Australopithecus afarensis* (Southern Ape from Afar), while the specimen herself quickly became known as "Lucy", named after the Beatles song "Lucy in the sky with diamonds" apparently a popular camp-fire favorite of Johanson and his team (it seems that the Beatles are big amongst anthropologists). Scientific facial reconstruction has demonstrated that she too (like Ardi before her) was no modern-day beauty (again, not a face you would want to wake up next to each morning), but the 3.2 million year old Lucy soon became a worldwide "celebrity", has been the focus of many exhibitions, and has even been on a lengthy tour of the U.S. (well the 40% of her found anyway).

Now our 2 *afarensis* relatives of Lucy who so helpfully left their footprints in the muddy ash at Laetoli 3.6 million years ago may well be romantically assumed to be a young couple out for a moonlit stroll, but, just like Lucy, we would be hard pressed to recognize them as being something close to a relative of ours. This was no Ken and Barbie. They would have likely been less than 5 feet tall, a brain about a quarter the size of modern humans, had long arms and short legs, had large teeth more suitable for their diet of nuts and berries, were covered in hair (we presume), a face with forward projecting jaws, and had little or no forehead. They may well have been a couple of super-models some 4 million years ago, but you wouldn't find them on the cover of any of today's fashion magazines or selling perfume on your TV.

But the footprints they so helpfully left at Laetoli clearly demonstrated that these early *hominids* walked upright as there is no associated knuckle impressions found that would have indicated an on-all-fours knuckle-walking approach. Analysis of the footprints also show that the owners of the feet did not have the mobile big toe you would expect to see if these were apes, rather they had an arched foot typical of modern humans, along with a clear human-like heel-to-toe action, and so the footprints were deemed almost indistinguishable from modern human footprints, but with the rather significant difference being that they were made almost 4 million years ago.

Before the discovery of the Laetoli footprints there was still much debate as to whether our ancestors first developed a larger brain (as the now exposed Piltdown man hoax had implied), or whether they just went ahead and adopted bipedalism long before any significant increase in grey-matter and associated cognitive thinking. But the discovery of these footprints, associated with the fossil evidenced small brained *Australopithecus afarensis* of that time, settled the argument once and for all, and proved that our early ancestors were fully bipedal long before the date of any evidence supporting an enlarged noggin, and long before the evidenced date that any stone tools were first being used (the presumed evidence for use of such an increased brain capacity). No evidence of stone tools, no matter how primitive, have ever been found alongside any *A. afarensis* specimens.

Thus it seems that those early ancestors of ours who stepped down from the safety of their arboreal home to make a new life on the ground standing up on two legs, were still a very long way away from being Albert Einstein (the evolutionary change that would see those currently small-brained "Forrest Gump" ancestors turn into large-brained tool-making stone-age rocket scientists was still at least a few million years away). Thus, the evidence does seem to rather support the fact that their decision to "step down and go vertical" was likely more "forced" upon them out of the pressing need for a daily square meal rather than as the result of some well-reasoned plan driven by a pioneering and adventurer spirit.

However, regardless of the reason, once the decision was taken it seems that those early ancestors of ours had inadvertently taken what is likely the single most important step on the evolutionary path to you and me. They were now out of the trees and had two feet firmly (well almost) planted on the ground.

12

A Few Side Experiments

Mary Leakey's find at Laetoli was not to be her only contribution to the ever growing story of mankind's evolution. In fact she was destined to become the matriarch of the greatest fossil hunting dynasty of all time, a kind of anthropological version of the Kennedy's. This "First family" of fossil hunters included her husband Louis, her son Richard, and her daughter-in-law Meave. Each has been responsible for unearthing pivotal finds that have proved critical to uncovering our seemingly complex evolutionary path.

Louis Leakey was born in 1903 the son of missionaries in East Africa where he was to grow up before going on to study anthropology at Cambridge University. While there he became convinced that the roots of mankind lay not in Europe as was the accepted belief at that time, but in Africa, and became obsessed that the best place to discover evidence of such roots was at a place called the Olduvai Gorge, a steep ravine about 30 miles long, in Northern Tanzania.

In the 1930's, by now already married with 2 children, Louis met Mary Nichol an unconventional free-spirited adventurer who had never gained a single academic qualification but whose love for adventure was matched only by what turned out to be their shared obsession of archaeology. Their mutual attraction (and hormones) were clearly too much for either of them and Leakey promptly left his wife and kids, married Mary, and the two of them then set off into the sunset to set up camp at the Olduvai Gorge where they both began a lifelong search for evidence of mankind's evolution. A search that was later to be taken up by their sons Jonathan, Philip, and Richard along with his equally fossil obsessed wife, Meave.

And it was Maeve Leakey who was to provide the scientific world with the first credible link between the primitive *hominids* we saw of 4.5 to 8 million years ago that first split from the chimpanzees, (our friends *Sahelanthropus tchadensis*, *Orrorin tugenensis*, and *Ardipithecus ramidus)*, and the *Australopithecines* of Lucy and her *A. afarensis* brothers and sisters, who appear to have roamed the savannahs of Africa between 3 to 4 million year ago.

In 1994, Maeve Leakey, excavating at a site near Lake Turkana in Kenya, uncovered fragments of what was clearly an early *hominid*, and included a lower jaw-bone which seemed to closely resemble that of a common chimpanzee, but whose teeth had a more human-like appearance (although presumably minus the cavities and bridge work). Taking note of the age of the fossil find, between 3.9 and 4.2 million years old, and the differences with the more recent *afarensis* fossil finds, she promptly assigned the find to a new species, *Australopithecus anamensis*, taking the name from the Turkana word "anam", meaning lake. Leakey quite logically argued that *anamensis* was very likely the direct predecessor of *A. afarensis*.

Further finds for *A. anamensis* were later made in Ethiopia, and thus placing the range of *A. anamensis'* stomping ground to be remarkably close to the discovery site of Ardi and his *Ardipithecus* friends, and dated to within a few hundred thousand years of when Ardi was apparently roaming the same savannah. This then in theory rather neatly provided the needed missing link between the early *Ardipithecus* species and *A. afarensis* who as we have seen (courtesy of those footprints in Laetoli) was clearly now walking upright, and who first appear on the scene around four million years ago.

So what then of Raymond Dart's Taung child and his (or her) people, *Australopithecus africanus*, who we now know were happily wandering around the southern end Africa between 2 and 3 million years ago? Well it's a logical step, and actually one that held some scientific support for quite some time, to pretty much just draw a straight evolutionary line leading from *Sahelanthropus* to *Orrorin* to *Ardipithecus*, then onto *A. aficanus* via the intermediary steps provided *by A. anamensis* and *A. afarensis*. We could then presumably just continue the same straight line, with each new *hominid* species "evolving" our line so far and eventually handing over the evolutionary torch to the next, itself then to continue the steady march directly on to you and me. Simple. Well, not quite.

However, it now appears that in a plot twist worthy of a Mexican Soap-opera, evolution was also conducting a few other little experiments in human evolution around the same time that *Australopithecus africanus* was happily roaming around Southern Africa. Our evolutionary path was about to become less of a convenient straight line and much more a series of experimental forks in the road, with most it seems eventually leading to evolutionary dead-ends.

Enter Robert Broom, a Scottish doctor who had been working in South Africa thru the 1930's, but who was also an enthusiastic amateur palaeontologist and who more importantly was a strong supporter of Dart's claims for *A. africanus* being an ancestor of *Homo sapiens,* even though science had up to that point dismissed Dart's claims for his "Southern ape-man" as no more than an "unfortunate looking" modern South African, all based purely on the findings driven by the soon to be discredited Piltdown "orang-panzee-man".

As such Broom began making his own excavations in South Africa to try to find additional specimens in support of Dart's claims. Broom was also known to be as eccentric as he was brilliant, obsessed with the benefits of sunlight he would often choose to excavate sites while stark naked (we have to hope and presume he didn't see the same need for vitamin D while consulting with his patients), and was often cited as stating that it was "spirits" that guided him when deciding on where to dig for fossils.

In 1938, Broom, now at the ripe old age of 70, while excavating at a site in Kromdraai (and hopefully not naked again given his age) not far from modern-day Johannesburg, discovered pieces of a skull along with some associated teeth, which he assumed would offer further specimen samples for Dart's *Australopithecus africanus*. But after further finds of similar looking specimens (presumably as directed by the "spirits"), Broom became convinced these may actually be a different species altogether. On closer examination the specimens indicated a somewhat brawnier "ape-man", with a head shaped more like a modern day gorilla, with a large projecting face, massive molar teeth, and a very significant ridge of bone running lengthwise along the top of the skull (known as the *sagittal crest* for those with scientific leanings) which is indicative of exceptionally strong jaw muscles usually found in animals that rely on a powerful chewing ability, indicating a diet based more on hard gritty foods such as nuts and tubers.

The find was clearly indicating a specimen that was much more than just a possible brawnier older brother of the Taung child on steroids, Broom it seems had stumbled upon a whole new type of *Australopithecine*, one which he duly named *Australopithecus robustus*, (for obvious reasons), and one which was eventually dated to live around 2 million years ago, and thus one which would have been roaming around Southern Africa seemingly vigorously chewing on nuts, berries, and tubers, at roughly the same time as *Australopithecus africanus*.

Broom did go on to find additional specimens of *A. africanus*, and his published findings over the course of the next 20 years were instrumental in gaining the eventual recognition that Dart's original find deserved. Broom's discovery of *A. robustus* along with his supporting work for *A. africanus* were remarkable, but probably most importantly of all, Broom provided clear evidence that human evolution was not based on the evolution of a single evolutionary line, but that for a significant period of time at least two types of early *hominid* were alive and thriving at the same place and same time.

For the first time science began to understand that the path of human evolution may in fact be one rich in diversity, and certainly one that was going to be more complex than originally thought. Clearly deciphering our evolutionary path was not going to be an easy "slam-dunk", the path was revealing itself as less of a stroll down a straight and narrow pathway, and becoming more and more like a blind-folded visit to a large and complex maze.

Then in 1959 Mary and Louis Leakey, still camped out in their beloved Olduvai Gorge, made a discovery that would complicate the picture even further. Digging around at a their latest site Mary unearthed a well-preserved skull and some teeth which dated as being around 1.75 million years old. The skull had a distinctly "robust" anatomy similar to that of A. *robustus* discovered by Broom some 20 years earlier, however they differed from Mr. Robustus in that the cheekbones were remarkably flared and the back teeth were huge, really huge (almost 4 times the size of modern humans) while the front teeth were tiny. The Leakey's initially named the species as *Zinjanthropus boisei* (Boise's East African man), after Zinj an ancient name for East Africa, while *boisei* gave a nod to Charles Boise who had been the Leakey's financial benefactor for a while. However, convention later led to a tactical renaming, and as such the species soon became officially known as *Australopithecus boisei*.

This early *hominid* was clearly as much of a brute as *A. robustus* with a highly specialized skull and jaw filled with tombstone-sized teeth, and as such it soon became referred to as the "Nutcracker Man". If alive today, even though analysis of subsequent fossil finds for the species indicate they likely only weighed around 100 pounds and stood no more than about four and a half feet tall, Nutcracker man would likely be making his living as a vegetarian night-club bouncer rather than as a surgeon, to paraphrase Theodore Roosevelt, "I could carve a better man out of a banana". Regardless, it appears that Mr. Nutcracker successfully inhabited his east African homeland for well over a million years dating to roughly somewhere between

2.5 to 1.2 million years ago, and so if nothing else he was clearly a survivor.

The ever argumentative academics that care about these things spend a great deal of their time (something they apparently seem to have a great deal of to spare) debating whether Broom's *A. robustus* and the Leakey's *A. boisei* should be classed as Australopithecines at all. They argue that their specialized features imply they are a quite distinct species on their own, as such they have looked to place them in their own *genus* (evolutionary related species) named *Paranthropus* from the Greek words *para* meaning "beside", and *anthropus* meaning "human".

And in a quite remarkable concession given the history of the field, those that continued to instead argue that both do rightfully belong within the *genus Australopithecus,* conceded that they would instead be classed as "robust" *australopiths*, while the others of the *genus* (*africanus, afarensis*) would be differentiated as "gracile" *australopiths*. However, at the moment the *Paranthropus* camp seems to hold sway, and thus Mr. Nutcracker and his South African cousin are now more commonly known in scientific circles as *Paranthropus robustus* and *Paranthropus boisei*. These things are important it seems in the world of palaeoanthropology.

Then, just to complicate the picture just a little more, there is the strange case of *Australopithecus aethiopicus (*first discovered in 1967 by some more of those excitable French who were by now running around southern Ethiopia still amusing themselves with more unfathomable Gaelic customs), but with the first significant find not made until some 20 years later in West Turkana Kenya, and which consisted of a fossilized skull that was estimated to be around 2.5 million years old. The species is considered by many to be a likely direct descendant of *A. afarensis*, while others also point to its "robust" similarities with *P. boisei* and *P. robustus* and thus see it as a possible common ancestor of both, (in which case as we have seen, to appease the academics it should more rightfully be called *Paranthropus aethiopicus*).

The list of identified species seen as appearing "somewhere" in our ancestral line, and seemingly also alive and kicking in various parts of Africa around 2 million years ago, is rounded off with two other proposed species. *Australopithecus garhi*, about which we seem to know very little other than someone found some fossils of a human-like creature somewhere in Ethiopia, and decided to name it after a word in the local dialect meaning "surprise" (which it is was to everyone that this was considered a new species at all based upon just this single find), and *Australopithecus sediba* whose claim as a unique and separate species is based purely on the fossil remains of six skeletons discovered at the bottom of a deep hole somewhere in South Africa where they all seem to have collectively fallen to their death, (apparently either in the first recorded incident of a mass murder, mass suicide, or collective stupidity).

Today we are used to being the only "human" species on the planet. We have in today's modern chimps and gorillas some very distant cousins roaming around the various jungles of the world, but as "modern humans" we have never had to share our planet with other "versions" of our human line. However, this has only really been the case for the last 20,000 years or so, a tiny fraction of the 5 million years that make up our distinct evolutionary history.

We can now clearly see that at some point between two and four million years ago there may have been as many as half a dozen different *hominid* species happily running around various parts of Africa at the same time, and possibly even more yet to be discovered. It was a veritable smorgasbord of human evolutionary branches.

However, it seems that very shortly for all but one of those species the evolutionary Fat-Lady was getting ready to sing. Evolution can indeed be cruel. But for one, it was fated not only to avoid the unfortunate consequences of evolution's ruthless selection process, but it would also go on to give rise to a whole new and quite remarkable family of *hominids*. No one really knows quite how it happened, and indeed quite what their relationship was with those

Australopithecines of 2 million years ago, but at some point around this time the *Australopithecines* would slowly and mysteriously begin to disappear from the savannahs of Africa, and be replaced by a newly evolving species, a species so unique it would need to be assigned its own evolutionary group, the *genus Homo,* the *genus* of all species of what we would recognize as being truly man, the *genus* of you and me.

13

From Ape to Man

As you would expect, there is a great deal of dispute around exactly which branch of our early ancestors (the *Australopithecine* family) it was that was spared evolution's trap-door and went on to further our evolutionary line from *genus Australopithecine* to *Homo* (from merely an "ape-like" man to fully fledged "human"). Separate cases have been made for *A. garhi, A. sediba, and A. africanus,* some even proposing *P. aethiopicus,* and in all honesty we may just as well lump in there aliens from another planet, because in truth our clever paleoanthropologist types really don't have a clue, and all this is notwithstanding the distinct possibility it could even have been some as-yet undiscovered species that has to-date slipped under their fossil-radar.

But why would one fortunate *Australopithecine* family (whichever one it was, and assuming we are ignoring some travelling inter-galactic evolutionary wild-card) become the winner of evolution's lucky draw and go on to kick-start the *Homo* evolutionary line, while the other handful of in-flight evolutionary branches which to be fair had seemingly survived quite happily for at least the last few million years, suddenly now find themselves being moved over to the "soon to be cancelled" list of evolutionary experiments?

Were the remaining *Australopithecines* herded off to form the lost city of Atlantis, were they all mysteriously transported away by visiting aliens for later study, or did they collectively evolve into lawyers and politicians which many believe are a distinctly different species from the rest of today's *Homo sapiens* anyway? Well, less dramatically, the answer it seems may well lay once again with the

Earth's climate, the same natural force that had drawn the ancestors of the *Australopithecines* down from the trees some 3 or 4 million years earlier.

It seems that around this period (about two million years ago) the Earth's climate was again going through a period of intense fluctuation, with periods of alternating warming and cooling causing ice-caps to advance then retreat, long periods of drought followed by long periods of cold, and over a period of several thousand years causing chaos to local ecosystems. The associated changes in temperature and food supply would have severely tested each of those early ancestors' ability to survive, and thus putting something of a strain on each of the poor old *Australopithecine* evolutionary lines. These were trying times for any species, but particularly for any species that harboured some loftier evolutionary ambitions.

To survive such upheavals in their environment each of these *hominid* species would have needed two very important things. Firstly they would have needed a very healthy dose of luck (something rather unfortunately out of their control), and secondly they would have needed to learn how to improvise and become more versatile, by learning to adapt their behaviours to such an ever-changing environment (something just slightly more under their own control). If you're living off a diet of nuts and berries, and the nuts and berries start to disappear, you either need a dietary "Plan-B", or you need to start working on your bucket-list.

It appears that faced with such challenges, one particular species of *Australopithecine* responded in such a way that it would not only secure its continued success as an active evolutionary line, but the way in which it responded spurred such a level of evolutionary growth that it was soon to become that founding member of the new family *Homo*. As for the other *Australopithecine* species, it seems they all chose to rather work on their bucket lists.

And what was this great new adaptation? Well it seems that presumably fed up with a diet of just nuts and berries, this "outside of the box" thinking species decided to pick up a flint rock one fine day and crudely fashioned it into what became the first stone tool, they then used this crude tool to scavenge and hunt for alternative food sources, including meat, and thus increasing the dietary options on their daily menu, and more importantly their chances of survival. This new breed of *hominid* not only became a tool-maker, but now also, at least in part, became a carnivore by adding meat to his daily two veg, and most importantly of all he had found a way to adapt and survive. Suddenly, now armed with a sharpened piece of flint, an intellectual bridge was crossed, and in doing so we were now a giant step closer to what was to become you and me. Stone-age man it seems, had been born.

So, just as Ardi and her buddies responded to climatic change by stepping down from their arboreal home onto the savannah several millions years earlier, those early ancestors of ours who were destined to become the first members of the "*Homo* family" similarly responded by adapting to their environment by learning to utilize stone to fashion rudimentary tools to overcome their own particular challenges. And for those who found themselves in evolutionary lines that failed to adapt, well unfortunately they would soon just find themselves on the wrong side of evolution's ledger-book, effectively marking the point at which our own evolutionary line, *Homo*, split from the soon to disappear *Australopithecines*.

And not surprisingly, it was to be a member of the Leakey dynasty, Jonathan Leakey, the eldest son of Louis and Mary who by now was joining his ever enthusiastic parents on their numerous expeditions in the Olduvai Gorge, who would be the first to find significant fossil evidence of what was believed to be that first early tool-making ancestor in the *Homo* line.

On an expedition at the Olduvai Gorge in May 1960, Jonathan Leakey (at that time only aged 19 and as such a mere fossil hunting "newbie"), unearthed several pieces of a *hominid* skull. Further digging around in the dirt and dust revealed the lower jaw, some teeth, and fragments of a hand, along with fossil remains from other similar specimens. But what caught Team-Leakey's interest about these specimens was that despite being dated to around 1.8 million years ago, the hand showed remarkable similarities to that of modern humans, but most strikingly the brain was clearly larger than that of any of the more cerebrally challenged *Australopithecines* found to-date. Clever work on the chalkboard seemed to indicate that these specimens had a brain capacity of around 600 ml which was almost 50% larger than that of your average *Australopithecine*, (but still significantly smaller than that of a modern human who weighs in with a hefty brain capacity on average of around 1,500 ml). These new *hominids*, whoever they were, would clearly all be considered rocket scientists when compared to their nut and berry chewing cousins.

But most remarkable of all was that lying alongside the fossils, the Leakey's found what appeared to be stone tools (admittedly very basic), along with animal bones that showed clear markings of having being scrapped clean. Not only did these new *hominids* have more dextrous hands and larger brains they were clearly using them, and using them to fashion tools to scavenge and strip the meat from their prey. They were clearly using their larger brains to find ways to adapt and survive.

Science bods whose job it is to sit around a table and define such things tell us that the key evolutionary changes between any new species who wish to qualify as being counted under the *genus Homo* rather than under the "still more ape than man" *Australopithecines* are a significant increase in brain size and associated signs of increased intelligence. Which can all be loosely translated to mean the manufacture and use of stone tools (it's not enough just to have a big brain it seems you need to show you can put it to some use). As such, having seemingly ticked all the prescribed boxes, Louis Leakey was quick to jump at the opportunity son No. 1 had given him to proudly announce that they had discovered the first species in the *Homo* evolutionary line. Thus, in 1964 having gathered what he believed was sufficient supporting evidence, he promptly announced to the world the new species, one he had proudly named as *Homo habilis* ("Handy Man").

Leakey's *Homo habilis* was, as expected, met less with fanfares and trumpets and more with skepticism and controversy around its presumed lofty position in the human family tree. Certain "rival" paleoanthropologists pointed to the fact that subsequent fossil finds attributed to *Homo habilis* showed it to be more Forrest Gump than Albert Einstein, leading many to believe it still didn't quite make the grade to be counted under the family *Homo*, and so should more correctly be classified as *Australopithecus habilis*.

Regardless of naming conventions and controversy, *Homo habilis* (barring a few die-hard naysayers, the currently accepted species name) had arrived on the scene sporting a skull filled with a state-of-the-art brain and hands wielding stone tools (the must-have accessory of the day), and it appears that as a species he was successfully using both to adapt to his ever changing environment somewhere between 2.5 and 1.5 million years ago.

14

Off to See the World

No one has been able to definitively say why all of a sudden in one evolutionary branch of our ancestral tree their brain started to evolve to be larger and far more complex than the model that had served all other branches pretty much unchanged for millions of years (maybe our own brains haven't grown big enough themselves yet to figure it out).

There is of course a big cost to having a larger brain, a big brain needs energy, and a lot of it. In modern humans (well most anyway) our brain is only 2% of our body weight, but it uses 20% of our energy, the brain is clearly a very needy piece of equipment. This means we need to consume a lot of food to fuel such energy demands, not so much of a problem for us modern-day city dwellers who can always just pop down to the supermarket to stock up when the food cupboard starts to look a little bare, but for *Homo habilis* who clearly wouldn't have had the luxury of a local Safeway, this would have been something of a challenge. To support his revolutionary brain, Mr. Habilis would have had to consume far more calories each day than his *Australopithecine* cousins who could seemingly still get by just grazing on nuts and berries. *Homo habilis* with his growing brain and intellect would have needed richer sources of energy, a steady diet of just nuts and berries was just no longer going to cut it, no, what he needed was some protein-rich, fresh of the bone, red meat.

Many believe that it was the growth in brain size that gave *Homo habilis* the cognitive wherewithal to fashion stone tools, which then opened the door to a wider variety of dietary options including energy-rich meat, and which helped support an evolving and growing brain, which then in turn enabled better adaptations to changing environments, meaning better tools, and so on and so on, effectively creating a cycle of "growth". The reality however, is that science really just doesn't know what kick-started such evolutionary changes in the brain in the first place. It may just simply have been a genetic accident, or maybe an individual ancestor who suffered a personally unfortunate but evolutionary beneficial bang on the head that had triggered the process. But whatever it was it clearly began an evolutionary experiment that was to ultimately lead to you and I.

However, *Homo habilis* was not the first species to have been found that was to be linked to the genus *Homo*. As long ago as 1887 a certain Eugene Dubois, a military surgeon working in Southeast Asia who had a particular interest in human evolution and in particular in finding evidence of modern man's early ancestors, set off to Sumatra in pursuit of his goal armed only with his enthusiasm, a handful of convict laborers (although enthusiastic, Dubois certainly didn't intend to do any actual digging himself), and in the knowledge that the area seemed to be full of caves, which to Dubois seemed a reasonable starting point. These early ancestors had to live somewhere, and some fully waterproof, protected on three sides, "spacious, well-appointed, and convenient for all local amenities", ready-made holes in the rock, seemed as good a place as any.

By 1891, having failed to find the fossil evidence he was searching for Dubois moved onto the island of Java where he began searching along the Solo River at a place called Trinil. It was here that he (or rather his convict minions) found fossil remains of a skullcap, a leg bone and a few teeth, all dated at the time to be around 700,000 years old. Although only a partial skull it showed that the original owner had a brain somewhat larger than any modern ape, while Dubois (presumably applying some liberal assumptions that remarkably modern scientific techniques have now shown to be quite correct) deduced from that single leg bone alone that the said owner had also walked upright. Dubois named his newly found species *Pithecanthropus erectus*, meaning "upright ape-man", and promptly declared it to be the "missing link" between ape and man. Unfortunately for Dubois no one wanted to listen.

To the European scientists of the day Dubois was a mere amateur with a handful of dubious "foreign" bones, who was also sat on the wrong side of the world, and so his find was dismissed as a mere a fossilized ape, and duly ignored (this of course would not the last time that scientific snobbery and nationalism would indirectly apply the brakes to the progress of our understanding of human evolution). It's fair to say that Dubois was not entirely pleased with how his new "discovery" had been received, and so naturally with his "enthusiasm" for his hobby now somewhat dampened, and his convict labourers presumably safely back under lock and key, he promptly disappeared from public view, never to be heard of again.

Then in 1927 a Canadian named Davidson Black, who was at the time an anatomist at the then named Peking Medical College in China, but who was also an enthusiastic fossil hunter, was directed to a site known as Dragon Bone Hill at Zhoukoudian not far from Beijing, which was apparently famous amongst local quarrymen for fossils. At first Black unearthed just fossilized molars but it was enough for him to declare the discovery of a new species in the human evolutionary line (rather presumptuously it was felt by some given that all he had for evidence was a handful of teeth), one he named *Sinanthropus pekinensis* (China Man from Peking) which very quickly became known simply as *Peking Man*. Further excavations over the following years eventually uncovered over 200 human fossils from what is thought to be around 40 individual specimens, and included six nearly complete skullcaps, and all dated at the time to be around 500,000 years old.

Unfortunately, excavations had to come to an abrupt end in 1937 when the Japanese rather rudely invaded China, and in 1941 the fossils were hastily packed to be moved to the USA for safekeeping until after the war. However the fossils mysteriously vanished on route to port, and like the disgruntled Eugene Dubois fifty or so years earlier, have never been seen or heard of since.

However, the quantity of the find at Dragon Bone Hill gave belated credence to Dubois' find, both seemingly indicating a bipedal species of large brain capacity and distinctly human-like features, and all seemingly running around various parts of Asia some 1.5 million to 200,000 years ago. Indeed further finds associated with *Peking man*, the evidence of fire and tool usage, clearly indicated a very human-like creature and possibly a direct ancestor of modern humans. As such, both Dubois' and Black's species were renamed, and today we know them collectively as the earliest examples of our not-so-distant cousins, *Homo erectus* (Upright Man). But *H. erectus* was to remain a mysterious and slightly murky species, being defined only by the handful of fragmented skulls along with some random teeth and bones, all gleaned from Java, and Black's now rather inconveniently "misplaced" casualties of war.

Now fast-forward almost 40 years or so, and Richard Leakey has assumed the role of head-honcho of the Leakey dynasty after the death of his father Louis in 1972, and was now dutifully carrying on the family business around Lake Turkana in Kenya. And it was there in 1984 that Leakey and his team were to stumble across one of the most significant finds in the study of human evolution, the *Turkana boy*.

When all the dust had settled, what Leakey had unearthed was over 100 fossilized bones that after careful study and some reconstructive jiggery-pokery represented a near complete skeleton of a boy aged around 12 years old, who was a quite surprising 5 feet 3 inches tall, and dated to be from around 1.5 million years ago. But what was even more surprising was the very human-like nature of the skeleton. When living, analysis showed that the boy still had something of an ape-like sloping forehead, and an absence of any real chin, but he also showed some very "human" characteristics.

For a start the *Turkana boy* had a noticeably bigger brain size (around 880 cc) compared to either the *Australopithecines* or even *Homo habilis*, the arms and legs were more proportioned, and in life he clearly had a very human-like protruding nose rather than the open flat ape-like nose seen on earlier *hominids*. He had also evolved a relatively tall and slim stature, presumably in response to the need to efficiently dissipate body-heat, a key adaptation for any species wishing to survive on the open savannahs of Africa (elephants of course are the obvious exception, but then they have ears the size of Egyptian fans they can flap around to keep cool).

It is also believed that Turkana boy had also lost most of the body hair that had been a trademark of his ancestral line (although the science that helped draw this conclusion remains murky at best), and as such he was believed to likely be dark-skinned as a protection against the sun (although this is also another assumption from the same school of murky science). Some clever science stuff based on the shape of the cranium also led to the belief that he was starting to evolve the area of the brain used in speech, and so likely he was capable of some very basic vocalization, and thus the beginnings of speech (unfortunately, it seems you can't excavate speech, and so this conclusion is also based on some more of that rather murky science stuff). Regardless, it seems that what Richard Leakey had stumbled across was the first *Homo erectus* specimen found in Africa.

Evidence found alongside this and other later *H. erectus* finds show that as a species they were the first to use fire, the first to collectively live in small hunter-gatherer groups, and were the first to fashion more complex tools. And in yet another remarkable find at Lake Turkana, fossil evidence showed a 1.5 million year old woman whose bones had been severely deformed by what was identified as hyper-vitaminosis A (vitamin A poisoning to you and me), an extremely painful condition. What was remarkable about this particular find was that this indicated (other than the woman clearly endured a very painful end to her life) *H. erectus* was very clearly now a meat eater (such poisoning can only come from eating the liver of a carnivore), but also that given the advanced state of her condition, the woman had lived with this incapacitating condition for quite some time, and so clearly must have been cared for by others of her group for a significant period. *H. erectus* it seems were caring for their sick. This was clearly a species that was no longer "ape", it was a species very much on the way to becoming "human".

And as evidenced by these fossil finds in both Africa and Asia, and subsequent finds that were soon to start turning up in various parts of Europe, *H. erectus* had also demonstrated that other very "human" trait, the urge to migrate and colonize other parts of the world, likely in search of better climates and more abundant food supplies, and presumably searching for the "promised land". He had clearly been on the move, and not just around the proverbial block, *H. erectus* had gone global.

But now our paleoanthropologist friends, who were by now locked away in darkened rooms feverishly trying to piece all this together, were soon to be faced with another unexpected problem. On paper it was all starting to look fairly straightforward, with *H. erectus* first appearing on the scene around 2 million years ago in Eastern Africa, evolving either directly from *Australopithecine* stock, or maybe even *Homo habilis* as some of the more controversial science bods have suggested (despite the fossil record seeming to indicate they actually co-existed in Eastern Africa for quite a period of time). The evidence then (courtesy of Dubois and Black) seemed to point to sometime around a million years ago Mr. Erectus got itchy feet and set off on his travels, slowly spreading out of Africa, and eventually into Asia where the fossil records seem to have shown him landing at Asia Border Control somewhere around 700,000 years ago.

But a problem was soon to surface. The dating of the fossil finds made in Asia by Dubois and Black all those years ago had been calculated based upon some rather archaic dating techniques that were now almost 100 years out of date and so were likely about as reliable as a politician's election promise. The problem was compounded by the fact that they were also based on some rather "dubious" data in the first place, seemingly collected not by slick fossil-hunting professionals but in the case of the Java fossils at least, by a mixture of convicts and farmers who have never been known as the most reliable sources for collecting accurate scientific data at the best of times.

And so a re-examination of some of the fossil finds from Asia, now based upon far more reliable "21st century" dating techniques, was about to drop a paleontological bomb-shell on the unsuspecting bods still smugly admiring their "straightforward" theory for *H. erectus'* evolution and world travels. What the revised dating indicated was that the fossil finds in Asia were at least 800,000 years older than first thought. This caused a huge problem, because if the new dates were right *H. erectus* seems to have made land in Asia around the same time he was still apparently emerging back in Africa. His slow spread out from his African homeland was now looking to be much more like a full on sprint.

This surprising news even led some overly-optimistic types to suggest that *H. erectus* hadn't initially evolved in Africa at all, rather he had evolved first in Asia and then spread out across the world from his Asian homeland. However, the fact that not one single piece of *hominid* evidence has ever been found anywhere in Asia that pre-dates *H. erectus*, was a slight stumbling block to the theory, particularly given that Asian born Mr. Erectus would have presumably had to evolve out of something rather than just miraculously appearing out of the blue one sunny Asian day with his flip-flops and chopsticks .

So it now appears that after emerging in Africa some 2 million years ago, H. erectus likely made land in Asia within the space of a couple of hundred thousand years or so. Quite a rate of migration, particularly given the obstacles they would have needed to overcome such as mountain ranges, some sizable bodies of water, and all presumably with no form of motorized transport. Not impossible, but certainly a great deal quicker than the science bods had first thought.

And curiously, the discovery that Mr. Erectus and his caravan of friends made their travel plans so quickly after first appearing on the savannahs of Africa, may well have also resolved one other little conundrum that had perplexed our paleoanthropologist friends for a while. As we have seen the older fossil evidence for *H. erectus*, dated from around 2 million years ago, had indicated the usage of stone tools, however these were clearly very rudimentary in nature. But it appears that the stone tools being found alongside Mr. Erectus in Africa some 500,000 years or so later indicated that he had by this time developed a far more advanced type of stone tool making technology. These more complex stone tools were more varied in nature, offering a greater range of application, and were clearly made in a far more controlled and precise fashion.

But here was the conundrum. There is almost no evidence at all to suggest that once settled in Asia, *H. erectus* was using stone tools anywhere near as sophisticated as those being found with their African cousins of the same period. They appeared to almost exclusively use the basic flint tools that back in Africa they had abandoned in favour of the more sophisticated tool making technology some half a million years earlier.

Perplexed anthropologists were left scratching their heads to explain why in Asia it appeared that *H. erectus,* right up to the point when they completely disappear from the fossil record as a species about 300,000 years ago, was almost exclusively using the stone-age equivalent of a horse-drawn plough, while back in Africa their contemporaries were now using a fully-motorized tractor. Apparently, when Mr. Erectus first left Africa to set off on his travels he left in such a hurry that he must have forgotten to pack his new stone tools, either that or his luggage was somehow lost somewhere along the way by the freight carrier.

However, the new dating of the fossil evidence from Asia now pointed to a far simpler answer. If, as the revised dates for the fossil evidence in Asia now indicated, *H. erectus* had indeed set forth on his travels east around 1.8 million years ago, he would have done so some 300,000 years or so before those left behind went on to discover any new and more advanced tool making technology. Thus, Mr. Erectus had arrived in Asia with knowledge of only the old, rudimentary, tool making technology, and for some reason never went on to evolve this any further. It seems he had just left Africa so early that he was blissfully unaware that their homeland contemporaries were now using a "tractor" while he manfully struggled on with his rudimentary "plough".

And just to add some additional spice to the usual conflicting interpretations that seem the norm in the science of human evolution, it is this gap in their tool making technologies, fuelled by some perceived physical differences between the relative specimens, that have led a section of the anthropological world to believe that the *Homo erectus* found in Africa (the now non-hairy, dark-skinned Turkana boy and his people), were actually a separate species altogether, believing that they should be more correctly named as *Homo ergaster*, ("workman"). They have further proposed that *H. ergaster*, having migrated out of Africa into Asia, branched out into a new distinct Asian species, *H. erectus*. Although *Homo ergaster* as a species has gained a degree of acceptance across the scientific world, most still usually only refer to the 2 species as only distinct African and Asian populations of the larger *H. erectus* species.

So then, by sometime around a million years ago, it seems that *Homo erectus* (along with *Homo ergaster* should you prescribe to that school of thinking) had evolved into a very human-like being who had successfully set up home across most of the "Old world". And thus over the course of several million years of evolution has taken us from tree-dwelling apes through to the first nervous upright steps of the *Australopithecines,* and then on to a tool-making, hunter-gatherer who cared for his sick mom, who could speak (just), had become tanned and distinctly less hairy, and who had rather quickly gone on to conquer the world.

There had been many "experimental" dead-ends in what appears to be the increasingly complex evolutionary tree that had got us to this point, but unlike the fate of the unfortunate individuals in those particular ancestral lines, the destiny of *H. erectus* was to be quite different. *Homo erectus* was destined to evolve. And as our next chapter in this story will show, exactly what *Homo erectus* evolved into, and where and when this happened, is the subject of just as much debate and contention amongst the scientific community as anything we have seen so far.

15

European Tourists

Much of what our science boffins now believe to be a credible evolutionary path to explain how mankind went from a tree-dwelling ape-like creature to be our planet's intellectual masters has seemingly been pieced together in the face of limited fossil evidence, conflicting "best fit", nationalist agendas, false trails, some rather clever hoaxes, self-promotion and bickering, and quite frankly a great deal of guesswork. Thus, it should come as no big surprise that such scientific theories will find at least a few willing opponents. And in this respect the loudest voice in the room comes from that ever solemn and devoted group, the *Creationists*.

Creationism is the religious belief that everything, including life, the Universe, planet Earth, mankind, the electric guitar, and the bikini, are all directly or indirectly the creation of some supernatural being, a God. A God who created mankind on either the last day of a frantic six day creation-marathon, by some form of evolutionary divine intervention, or even planting man on planet Earth fully formed jumping out of some "higher state of consciousness", all depending on which particular version of "God" you choose to kneel to.

Creationists will generally take a literal view of their chosen religion's teachings, and with scientific wisdom open to so much interpretation and conjecture, it should come as little surprise that even today, over 150 years after Charles Darwin first introduced us to the idea of "man from ape" evolution, it appears that almost half of all Americans still believe that God created the Universe, the Earth, mankind, and just about anything else you care to think of, all out of some glorious creation cookie-jar, and supposedly all within the last 10,000 years or so. To the Reverend C. D. Light and his flock, the Bible is the written word of God, and as such all its teachings must therefore be both historically and factually true, end of story, and thus their response to any conflicting scientific theory is usually to just put their fingers firmly in their ears and loudly hum hymns of redemption to themselves.

Now in the past even the most mentally gifted among our ancestors have been guilty of supporting beliefs that were subsequently proven somewhat left of the mark by scientific fact. A belief that the world is flat, that the sun orbits the Earth, that drilling a hole in your head is the correct cure for a headache, are all good examples. However, science, empirical evidence, and painful experience, have eventually provided clarity on such matters, such that we are all now happy to travel beyond the visible horizon to consult a doctor, all without the fear of either dropping off the edge of the world or coming home again with a rather large and unattractive hole in the side of our head. But when we consider our scientific theories around human evolution, to some extent they still struggle to gain acceptance amongst the wider population, particularly in the more fervent religious circles (as the 100+ million God-fearing Americans who still believe we are all descended from Adam and Eve rather than Lucy and Ardi, will attest).

Unfortunately, as a collective science, paleoanthropology (the study of human origins) relies as much on luck to progress and solidify its theories as it does on solid scientific work on the chalkboard. Unfortunately, you have to find the evidence first before you can start to study and interpret it, and as we have seen fossil evidence of our early ancestors are unfortunately not to be conveniently found under every rock you turn over.

Now, as more and more fossil evidence presumably does come to light and scientific techniques continue to improve around fossil dating and genetic analysis, the field of paleoanthropology, and its associated theories, will hopefully edge closer to empirical truth, and thus force more general acceptance (as long as we can persuade the non-believers to take their fingers out of their ears long enough to listen). But while each new fossil find continues to throw up potentially conflicting rather than supporting evidence, or our palaeontologist fraternity continue to bicker and argue amongst themselves around differing interpretations, then the door will always remain open for groups such as the *creationists* to question, deny, and raise doubt, around the validity of any scientific theories of human evolution (until someone had actually sailed around the world and ended up back where they started, certain sections of the population would always hold onto a belief that the world was indeed a pancake, not a ball).

And there is no better example as to the confusion, varying interpretations, and down-right conflict that exists in the hunt for a true understanding of our human evolution, than that which exists around *Homo habilis*. The problem it seems with Mr. Habilis is that there is a lack of clear evidence in support of them as a definitive species with very few *H. habilis* fossils ever having been found, we're talking a handful at most (likely there are more people living in the house next door to you than the total number of specimens found to-date that represent the whole *Homo habilis* species).

To make matters worse, it seems that the specimens that have been found to date and attributed to this species vary widely in their features leading to a real mishmash of supposedly "species defining" features. In short, the fossils attributed to the world's first "Handy man" just don't seem to make sense, and so not surprisingly, no two palaeontologists agree on the attributes that define the species, while many believe it should not even be seen as a separate species at all.

Firstly, fossil finds attributed to male *Homo habilis* seem to support the general theory that not unexpectedly indicate a species evolving more human-like features with an increasing brain size, but then some female *Homo habilis* fossils seem to indicate more ape-like features with an associated ape-like small brain, and thus not really showing any inclination to evolve at all, let alone be a credible transitionary species on the way to you and me.

Still others point to the fact that *H. habilis* fossils are often found with primitive stone tools, thus presumably indicating a link to our own direct tool-making lineage and thus making H. *habilis* a possible common ancestor of modern man. But then sitting on the other side of the classroom other science-types point to the fact that *hominid* fossils from several *Australopithecine* species, and even *Homo erectus*, have also been found from the same time periods, and so would have co-existed beside H. *habilis*, indicating that they could also just as easily be nothing more than another evolutionary dead-end and not a common ancestor of you and I at all. And in a final slap in the face for poor Mr. Habilis, some even see him as just a "bucket" species, one where those fossils that just don't seem to make sense to anyone, are all just conveniently lumped together in a kind of "when in doubt call it *Homo habilis*" approach.

All in all you do have to feel a little bit sorry for *Homo habilis*, as depending on which "well respected" corner of the paleoanthropological world you listen to, he is either a key transitionary species, an evolutionary dead-end, a miscellaneous bucket for waifs and strays, or not even a species at all.

It's all just another example of the scientific potential "train-wreck" that is our understanding of human evolutionary theory, and of course this all fuels the *creationist* fire which really just states that when it comes to human evolution, science just doesn't seem to have a clue. A little harsh, but to some extent also fair as even the most passionate of *evolutionists* will admit that our understanding of human evolution is still, well, evolving. So for now at least, Adam and Eve, and the Genesis story, will continue to remain the "truth" of human evolution for many around the world as they quietly hum their songs of redemption.

But one point on which there does seem to be at least a reasonable level of mutual scientific agreement is that by around 1.7 million years ago, via one of the several possible evolutionary paths that we saw unfold in the previous chapter, a very human-like species named

Homo erectus had emerged and set out from its African homeland and subsequently set up shop across most of the Old World including Asia and parts of Southern Europe. But as to what was to happen next to *Homo erectus* is where we unfortunately seem to plunge right back into the world of scientific head scratching and conjecture.

Firstly, in Asia as you will recall, *H. erectus* appears to have arrived in such a hurry that he completely missed out on any opportunity to benefit from his homeland contemporaries later evolution of a more advanced tool making technology, and thus he entered Asia probably out of breath and weary, and carrying with him what was soon to become a very out-dated stone-toolkit. Then, just to compound the situation, for some inexplicable reason he also appears not to have bothered to hone his own tool-making skills much beyond that point during the course of the next million and a half years or so that we see fossil evidence of his continued existence in the region.

Now whether we can point to this lack of technological advance and presumably the missed opportunities it brought in terms of any ability to better adapt and overcome their environmental challenges is debatable, but for whatever reason it seems that around 150,000 years ago we see *H. erectus* mysteriously disappear from the fossil record in Asia all together. If there is an evolutionary link or transitional species between the *H. erectus* of Asia and today's modern humans (*Homo sapiens*), its fossil evidence has yet to reveal itself. It seems that having been in such a rush to get to the promised land, *H. erectus* stuck around in Asia for around a million years, but in evolutionary terms effectively did very little, and eventually just seems to have faded away.

In Europe however a somewhat different story unfolded. It seems that those who waved goodbye to their Asian bound brothers and remained behind in Africa were not in quite so much of a rush to venture north into Europe. For a start it likely wasn't the most enticing of places unless you were partial to front-seat views of an ever shifting northern ice-cap, and slightly more chilly weather than

Africa was currently offering. Thus, by the time the European tourist industry had got itself organized enough to start to entice *H. erectus* north and once again out of Africa, he had already developed a far more advanced tool-making technology, while his brain had also started something of a growth spurt. Thus, the *H. erectus* that eventually packed his suitcase with thermal underwear and woolly socks and headed north into Europe was a somewhat more intelligent and sophisticated model than those that had sprinted off down the street for Asia a few hundred thousand years earlier.

In fact, human-like fossil remains found at the site of a huge cave known as Gran Dolina at Atapuerca, Spain, between 1994 and 1996 and dated to be around 780,000 years old, showed evidence of very "erectus-like" settlers, but whose remains indicated that they had a brain size of around 1000 cc, a significant increase when compared to the *H. erectus* running around parts of Asia and whose brain size seemed to have stalled at around a chimp-like 600-800 cc. This increase in brain size, along with evidence of stone tools indicating a more sophisticated tool-making technology, led to these early European settlers being proposed as a new species altogether, *Homo antecessor*, after the Latin word meaning "explorer" or "early settler", a name given to indicate that these were the first "human" population to get their passports stamped for Continental Europe.

As you would expect, debate raged as to whether *H. antecessor* now sitting comfortably in his newly furnished Gran Dolina cave had evolved first from *H. erectus* stock in Africa before moving north, or whether *H. erectus* had first departed on his northern walkabout and subsequently evolved into *H. antecessor*, or indeed whether Mr. Antecessor warrants his own species separate from *H. erectus* at all. But regardless, as other similar finds in the Atapuerca region of Spain dated to around 1.2 million years ago also indicated, it seems clear that "erectus-like" settlers had migrated from Africa into Southern Europe as early as 800,000 to 1.2 million years ago.

However, these first European tourists were not the first fossils of early humans to be found in the Atapuerca region. The most famous site in Atapuerca is the *Sima de los Huesos*, which in loose translation sounds like the title of a low-budget horror movie, "the pit of bones". The site itself is a bit of a horror story in its own right with a lengthy scramble through an intricate cave system just to get to the site entrance, which itself is located at the bottom of a 50 foot shaft (it seems our ancestors were not going to make it easy to unravel their evolutionary secrets, and thus potholing skills, or strangely "spelunking" as it's better known in N. America, are now also a basic requirement for today's multi-talented palaeontologist). The site however is a paleontological treasure trove giving up over 5,000 human bones of around 30 individual specimens since the site was first found in 1983, and all dated to be around 300,000 years old, and representing a remarkable three quarters of all human fossil remains found from the period 100,000 to 1.5 million years ago.

These fossils represented specimens of early humans living in the same area where *Homo antecessor* had set up home some half a million years earlier, and are assigned to the species *Homo heidelbergensis*, a species first discovered back in 1907 near Heidelberg in Germany, with subsequent finds in Greece, France, Italy and England. *H. heidelbergensis* was a true European settler. But there was to prove something different about *Homo heidelbergensis* that we had not seen in any previous species in our human ancestral line, and in a strange historical twist the evidence for this difference was to be found in Sussex, England, a corner of the world that as we have seen already had a rather dubious history with palaeontologists due to the earlier Piltdown man fiasco.

Boxgrove quarry in Sussex, England, is another of those rare sites that in a field where the find of a single fossilised molar after years of scratching around in the dirt is considered something of a success, was to prove another archaeological treasure trove. But Boxgrove's treasure was not to be fossilised human bones, but rather a vast cache of stone tools and weapons (a veritable arsenal) alongside numerous animal bones. This enormous haul was carefully amassed by archaeologist Mark Roberts over a period of 10 years between 1986 and 1996.

Numerous flint tools and weapons were unearthed alongside the remains of large animals, horses, rhinos, deer, bear, all of whose bones displayed evidence of wounds, cut marks and butchery, and all dated to be around 500,000 years old. Whoever used these weapons to kill and then butcher these large animals was clearly a very sophisticated species, and gave evidence that a critical change had taken place in man's activities, he was now showing signs of real intelligence. Evidence indicated that these animals had not been scavenged after they had died of natural causes or at the hands of some fearsome predator, rather they had been systematically hunted and killed by man using sophisticated stone weapons. Man was no longer a scavenger, he was a true hunter.

And hunting, particularly the hunting of large animals, means they needed to act collectively, in groups, plan and execute complex strategies, all of which means they must also have had a means to communicate such complex ideas effectively and efficiently and thus (to a certain degree at least) these Boxgrove hunters must have been able to talk.

These residents of S.E. England are now generally believed to be of *Homo heidelbergensis* stock dating as they are from roughly the same period as the specimens found in the "pit of bones" at the other end of Europe. Further evidence of their sophisticated hunting tools was also later found at Schoningen in Germany by Dr. Hartmut Thieme when he discovered 8 wooden throwing spears, all remarkably well preserved, and dated to be around 300,000 years old. The spears were expertly crafted and have been shown to be equivalent to today's modern tournament javelins in respect of their throwing qualities. *H. heidelbergensis* was clearly an intellectually superior species, (evidence shows that on average they had a brain volume weighing in at almost that of own), demonstrating highly developed technological skills, a complex social structure, and the cognitive and communication skills needed to execute sophisticated hunting strategies. Mr. Heidelbergensis then is where our human ancestors start to look and act human.

As you would expect from a field where it seems its professionals all have PhD's in "Advanced Argumentative Techniques" to go along with their fossil-hunting credentials, the evolutionary relationship between *H. heidelbergensis* and *H. antecessor* is the subject of much scientific debate. Some see *H. antecessor* as the evolutionary link between the African *H. erectus* who had been the first to set foot in Europe around 1.7 million years ago and the more evolved *H. heidelbergensis*, while others see them as really just one and the same and all part of the wider *H. heidelbergensis* European family that evolved from those early *H. erectus* travellers.

Regardless of the lineage, it seems that around 500,000 years ago Europe was the home to some very smart cookies, but what about those early ancestors who chose to stay behind in their African homeland when their more adventurous brothers and sisters had set off to conquer Asia and later Europe? What happened to *Homo erectus* in Africa?

The answer it seems had already been discovered way back in the early 1920's when a miner at a local zinc mine at Broken Hill in Northern Rhodesia (due to subsequent political upheavals, a place now known as Kabwe in Zambia) found a skull and upper jaw from 2 individuals dated to be around 300,000 to 125,000 years old. What was unique about the find is that the skulls indicated a brain capacity similar to that of modern humans (around 1,200 ml) with very modern-like features that indicated it was likely an intermediate species between *H. erectus* and modern-man. The species was initially named as *Homo rhodesiensis* and dubbed "Rhodesian Man", but most now believe Rhodesian man to be the first specimen found of *Homo heidelbergensis* in Africa.

So to quickly recap, it seems that around 1.7 million years ago somewhere in Africa, *H. erectus* suddenly got the urge to travel. One tour group set off immediately, not waiting to see what new tools could yet be developed that may have helped them on their travels, and, pausing only to check that their visas and health insurance was up to date, they quickly sprinted off into Asia where after some million or so years of little or no evolutionary growth they were to exit our story through one of evolutions trap-doors. A second touring party set off north from their African base-camp a little later to enjoy the somewhat chillier climes of Europe, but now were armed with the latest in tool-making technology, where they eventually evolved into the sophisticated hunter *Homo heidelbergensis*. Those that preferred the more familiar comforts of home stayed in Africa, where seemingly they too would independently evolve into a larger brained hunter who we also now recognise as a warmer climate, African model of *Homo heidelbergensis*.

Thus, around 300,000 years ago, *Homo heidelbergensis* likely stood at the point where 2 separate branches of mankind's evolution were independently unfolding. In Europe they would eventually evolve into the *Neanderthals*, while back in Africa their evolutionary path would lead to the emergence of a far different human species. The two species would not meet for about another 200,000 years, but when they did it would mean being consigned to the history books for one and the top of the evolutionary tree for the other.

16

Stone-Age Muggers

Very few races have suffered a worse press than the *Neanderthals*. Native Americans were vilified as nothing more than cut-throat savages with an over-fondness for wearing feathers by the 19th century settlers to the American mid-west who completely misunderstood either their culture or natural resistance to their attempts to "civilise" them, while several indigenous tribes of Central Africa were seen by Victorian-era Europeans as nothing more than cannibalistic and godless barbarians who would rather use the missionaries sent to "save" them as ingredients in their over-sized cooking pots. But even these historically embarrassing misconceptions were small potatoes when compared to the less than flattering press the poor *Neanderthals* received.

From their first discovery in 1856 by quarrymen in a cave in the Neander Valley near Dusseldorf, Germany, the *Neanderthals* as a species were for the best part of 100 years caricatured as nothing more than dim-witted, lumbering, fearsome, sub-human, savage brutes. Unfortunately, their physical features didn't necessarily help their cause either, with a face displaying a very prominent brow-ridge, low narrow forehead, large flared nose, a receding chin, with "murderous-looking" eyes, all sat on top of a squat, stocky and robust body, (and that was just the females). They were seen as merely hairy cavemen running around the countryside with big stone clubs, ripping up trees, grunting meaninglessly at each other, and looking for innocent victims to ambush, kill and then eat (and not necessarily in that order). Imagine a short, stocky, bad tempered wrestler with the brain of a 2 year old and the temperament of Attila the Hun, and you'll have a pretty good picture of the public perception of *Neanderthals* right up until the middle of the 20th century.

Yet rather clever modern scientific bods have now proven that these "muggers on steroids" actually had brains at least equivalent in size to our own (in many cases actually larger) and although due to the harsh conditions under which they had to survive they can never been seen as a peace-loving, tree-hugging, flower-power hippie people, we shall see that they were a far more sophisticated, intelligent, spiritually aware, and highly resourceful species than the one painted of them by science right up until the 1950's.

Part of the problem was that there was a desire amongst Victorian and Edwardian scientists to ensure that *Homo sapiens*, particularly those *Homo sapiens* from Western Europe, were placed firmly at the top of the evolutionary ladder. With the existence of *Homo neanderthalensis* being dated so close, or even overlapping, with our own emergence as a species, it suited the scientists of the time to paint the *Neanderthals* as nothing more than brutish, savage ape-men (when creating mankind in His own image, God had to be seen as clearly only having *Homo sapiens* in mind, He certainly could not be seen as tinkering with a possible "alternate model", that certainly would not do at all). Thus, a very firm line needed to be drawn between ourselves and any other "human" species found in the fossil record, and so each new fossil find of *H. neanderthalensis* was thus treated to a somewhat unsympathetic scientific analysis. A prime example being the discovery and subsequent analysis of the "Old Man of La Chapelle".

In 1908 three priests of all people discovered a complete (well almost) skeleton with very Neanderthal-like features buried in a pit at the site of a cave at La Chapelle-aux-Saints, Southern France. The bones of the "Old man of La Chapelle" as he became known, were passed to a certain Marcelin Boule, a French palaeontologist, for study and who went on to describe the skeleton in a paper published in 1911.

As expected, Boule's paper did the *Neanderthals* very few favours, painting a very clear picture of a brutish half-man half-ape, Stone-age mugger, all supported by what we now know to be rather less than sound scientific analysis. His intentional demonizing of the *Neanderthals* was very much in line with the needs of the scientific community of the day, and it would take many years for *Neanderthals* (and science) to overcome the image that Boule had created. Unfortunately, it seems that like unwarranted gossip, scientific "mud" however unfounded, still sticks.

Since their first discovery in 1856, palaeontologists have gone on to find more fossilized specimens of *Neanderthals* than any other species across the 8 million years that our particular evolutionary tree has been growing (outside of those of our own particular species of course), and many of those have been in excellent condition. But it was not until almost 100 years after that first discovery in the Neander Valley that science started to put away its prejudices and *H. neanderthalensis* began to be viewed with anything other than unbiased eyes, and so their true nature finally began to come to light and receive proper scientific recognition.

We now know that *H. neanderthalensis* first appeared in Europe around 250,000 years ago, but unlike their presumed immediate ancestors *H. heidelbergensis* who themselves were born of African *H. erectus* stock, they were shorter, stockier, heavier, and had larger brains which they clearly used to great effect as they managed to call what was at that time a very cold and unwelcoming Europe their home for the next 200,000 years, surviving it seems right up until about 30,000 years ago.

Europe during this period was no Club Med. resort, it was largely in the grip of an ice-age with large areas of Northern Europe at that time covered in vast sheets of ice. Much of the water was thus locked up in ice and glaciers, dropping ocean levels by up to 100 meters and so creating land bridges across the shallower seas. In such conditions, temperatures would have often been cold enough to put fingers, toes, and various other appendages at risk, with shorts and T-shirt days scarce and food even scarcer still.

Today it seems only Eskimos, polar explorers and penguins are foolish enough to choose to live in such conditions, and the *Neanderthals* clearly would have retreated to the less inhospitable southern reaches of the continent during the worst periods, but nonetheless, Mr. Neanderthal survived in such conditions for tens of thousands of years, and that's a lot of very cold, very harsh, European ice-age winters to survive without thermal underwear. In fact remains of *H. neanderthalensis* have been found right across Europe from England in the north to Italy in the south, and from the Iberian Peninsula in the west all the way across to Uzbekistan in the east. If nothing else, they were clearly survivors.

And that survival was in part due to the very physical attributes that Monsieur Boule had so conveniently used to define them as Stone Age half-ape, brutes. It seems that *H. neanderthalensis* had over time evolved rounded, stocky, muscular bodies, in an evolutionary response to the fiercely cold and rugged conditions in which they lived. Their wider, larger, noses enabled them to warm their breath enough to make it bearable, while such a "squat" stature enabled them to retain heat far more efficiently. Yes, they looked very intimidating, brutish even, but we now know this was just evolution adapting their original tall, slim, *H. erectus* stock built for the savannahs of Africa, into physical attributes more suited to a species eking out its existence in the harsh environment of what was effectively the giant fridge known as ice-age Europe. *Neanderthals* were built to survive, not to win beauty pageants.

In short, the *Neanderthals* were the first *hominids* to adapt fully to conditions in Europe, they were the first true Europeans. And this reassessment of *H. neanderthalensis* over the last 50 years or so has not only lead to a more "sympathetic" understanding of their physiques and looks, but we now have a far better picture of who the *Neanderthals* were as a people.

In 1992, at a limestone cave in Israel, an almost intact skeleton of a *Neanderthal* small child was unearthed. What was remarkable about the find was that the infant had clearly been laid in position with arms at their side, and with the jawbone of a deer placed over the infant's pelvis. The infant was dated to be about 60,000 years old, and had clearly been deliberately buried along with a ceremonial "offering" for the "afterlife". Over the years many examples of *Neanderthals* having intentionally buried their dead have been found. These were not the uncaring, "if it's dead then we may as well eat it" savages we had been led to believe, they cared enough for those close to them to ceremoniously bury them, (after politely letting them die in their own good time first, of course), and in doing so they were demonstrating love, respect and a level of spirituality.

And many of the *Neanderthal* fossil finds were also starting to help paint a picture of the type of life they were forced to lead. Almost every specimen found shows some degree of traumatic injury sustained during their lifetime, many bearing the scars of numerous injuries, broken bones, and lesions. They were clearly living a dangerous, rugged, violent, hand-to-mouth existence in harsh and unforgiving times. Such numerous and traumatic injuries likely indicate that they were forced to kill large game at close quarters, not easy at the best of times, but it also showed that as a species they were collectively hunting and were also caring for their sick and injured. Specimens indicate that despite the degree of trauma suffered, many *Neanderthals* lived long after the injuries were initially sustained, and given a presumed absence of any local ER services at the time, this is something quite unthinkable had their comrades not fed and cared for them as they convalesced. However, given their lifestyle, and despite their communal care, fossil analysis still indicates that it was a lucky *Neanderthal* indeed who lived much beyond the age of 30.

Given the conditions and a cold more akin to the freezer aisle in your local supermarket you don't need to be Sherlock Holmes to deduce that to survive the *Neanderthals* must have had at least a rudimentary understanding of fire, and had evolved some degree of "weather appropriate" clothing to supplement their admittedly already very hairy birthday suits. Indeed, there is evidence from several *Neanderthal* sites of their controlled use of fire, that they did indeed clothe themselves in the fur and skins from the animals they preyed on for food, and that they were making increasingly more complex tools. There is even some evidence to indicate that they made simple ornaments and necklaces for decoration. These were a caring, thoughtful people, who showed a remarkable intellect and a level of spirituality, all in the face of a harsh and challenging environment. The *Neanderthal* were now suddenly looking less like hairy football hooligans, and more like something of a Stone-age renaissance-man.

Modern scientific thought then is that Mr. Neanderthal was thus not that different from us in many respects, he may have looked somewhat intimidating to the more faint hearted amongst us, but clean him up, give him a shave and a haircut, put him in a business suit, and he may not quite pass for your local bank Manager but he would likely turn fewer heads on the street today than many devotees of some of today's more extreme "fashions". He looked like us, behaved like us, but was just not quite us.

17

The Rise of Homo sapiens

Of course, the obvious question to ask now is that if *Homo neanderthalensis*, Mr. Neanderthal, was indeed so like us, big brained, a natural survivor, and a species that had proved virtually indestructible in the face of the excesses of Europe's ice-age winters, why then did they somewhat abruptly disappear, with the last isolated pockets of evidence for their existence found clinging on at the southern-most tip of Spain about 30,000 years ago? Why are we not all sharing our towns and cities with clean-cut and clean-shaven *Neanderthals* today? Surely it would have taken nothing short of a nuclear Armageddon, or a meteor strike similar in size to that which abruptly saw off the dinosaurs, to force such an "indestructible" species to put on the concrete boots and jump off the side of the S.S. Evolution?

However, with Oppenheimer's Manhattan project not scheduled to begin for another 30,000 years, and with no evidence so far of any Texas-sized crater left behind by an inconveniently Earth-bound meteorite being found anywhere on continental Europe, we must assume some other force was at play that ultimately saw Mr. Neanderthal and his kind unceremoniously pruned from the human evolutionary tree. It was yet another mystery to solve in the story of human evolution, a story that was taking more twists and turns than solving the Rubik's cube blindfolded.

Well one theory that actually held sway for quite some time was that the *Neanderthals* didn't disappear at all, rather they merely followed evolution's natural course and had over time simply evolved into a new species, the *Neanderthals* had possibly just evolved into European modern man. This idea of a specific European evolutionary line from *H. neanderthalensis* to *Homo sapiens* was primarily supported by Franz Weidenreich, a German anthropologist who had made his name as one of the discoverers of the "Peking Man", *Homo erectus,* in China back in the 1920's. Weidenreich had published a paper in 1939 proposing an evolutionary sequence from Peking Man to the modern human line now found in Asia, a theory he later extended by proposing not only Dubois' "Java Man" as the natural ancestor of the modern Australian Aborigines, but also *Neanderthals* as the ancestor of modern Europeans.

Weidenreich believed that each of the Old World's inhabited regions (Africa, Asia, Europe, and Australia) all had their own distinct local line of evolution, all starting around 2 million years ago and which had all marched forward in roughly the same general direction, and all independently leading to the rise of *Homo sapiens* in each region. The theory, known as the *Multiregional hypothesis*, holds that when *Homo erectus* left his African homeland almost 2 million years ago in his great migration throughout the Old World, he went on to create discrete *H. erectus* populations in several distinct regions across the Old World, populations that had been independently evolving ever since, and all ultimately resulting in populations of *Homo sapiens* across each of these disparate regions. Hence, as Weidenreich explained, we can account for the physical differences we see in the regional characteristics of modern humans across the world, but with such "regional" diversity still remaining within the bounds of one "global" species, *Homo sapiens*.

At the time Weidenreich's theory naturally caused a great deal of uneasiness in the scientific and political world as it clearly implied distinct evolutionary lines and differences between regional instances of modern man across the world, and left the theory open to accusations of racism and the existence of deep divides between the world's populations. It was an uneasiness made even more awkward when you understand that Weidenreich himself was a German born Jew who had rather hastily fled to America in 1934 to avoid the growing Nazi racist persecution of the Jews.

The issue was somewhat mitigated when in the early 1980's Milford Wolpoff of the Michigan University, in support of the *Multiregional hypothesis* argued that although *Homo erectus* had indeed independently evolved into modern man across Africa, Asia, and Europe, but that there was also a significant degree of gene flow, interbreeding, genetic drift, and natural selection, between the regional lineages, all of which kept the overall "global" development of *Homo sapiens* in the same general direction. In this way, each regional evolutionary line gave birth to the same genetically distinct species, *Homo sapiens*, while still maintaining some key regionally unique adaptations.

Although this *Multiregional hypothesis* still retains a certain level of support even today (primarily by those with somewhat "other" agendas), it has now been mostly rejected by science given that it implies an improbable amount of independent and parallel evolution of *Homo erectus* through several separate evolutionary lines. The idea that evolution would take several geographically disperse instances of the same species, *Homo erectus*, and over the course of the exact same period of time, evolve each in exactly the same way, was too great of a stretch even for the usually creative-minds of today's paleoanthropologists (it's a bit like four artists in four different parts of the world, all independently creating the Mona Lisa at the exact same time, it could happen, but you shouldn't bet your house on it). In addition, scientists could see no fossil evidence supporting any evolutionary connection between *Neanderthals* and modern Europeans, or between Peking man and the Asian populations of today. The evidence just didn't support the theory.

So then if Mr. Neanderthal didn't just evolve into you and me, just what did happen to him? Well, to understand exactly what did happen to *H. neanderthalensis*, and indeed pick up the trail once more of the evolutionary line that would eventually lead to the emergence of our own unique species, *Homo sapiens*, we must head back to where our human ancestral line had first began around 7 million years ago. We need to jog back to the somewhat warmer climes of Africa where we had left Rhodesian Man, *Homo heidelbergensis* (African model), still running around in the warmth of Southern and Central Africa some 300,000 years ago.

You will remember we had noted earlier that around 1.7 million years ago our early *hominid* traveller, *Homo erectus*, had applied for new passports and the relevant travel visas, packed his suitcase and left his African homeland first for Asia (actually in quite the rush as you will recall) and then later for Europe where he had subsequently evolved into our very "human-like" Mr. Neanderthal via a brief evolutionary pause as the European model of *Homo heidelbergensis*. As for those who still preferred the warmth of the more tropical savannahs, and so stayed in the comfort of their African homeland, had by around 300,000 years ago themselves also evolved into a new species also now considered to be *H. heidelbergensis*.

But what had become of those that had preferred to stay on their more familiar African soil? Well the answer it seems was to once again be found by that ever industrious "first-family" of fossil hunters, the Leakey's. In 1967 Richard Leakey on an expedition to the Omo-Kibish region of Ethiopia unearthed a partial skull and skeleton of what appeared to be an early human-like *hominid* that was later dated to be around 130,000 years old. What was remarkable about the "Kibish Man" as it became known, was that later analysis of the skull indicated that it had a broad face, a high forehead, a prominent chin, and with a brain capacity on a par with modern humans. It seems that Kibish Man was not an early ancestor of ours at all, he was in fact clearly identifiable as a *Homo sapiens,* he was the same as you and I.

Thus it would now appear that while in Europe *H. heidelbergensis* was busy evolving into the "built for life in the freezer" *Neanderthals*, and in Asia *H. erectus* had seemingly given up on evolution all together, back in Africa Mr. Heidelberg had been slowly and quietly evolving into Mr. Kibish, he had evolved into modern man.

Similar finds of anatomically modern humans were later found in South Africa dated to be between 90,000 and 120,000 years old, along with some stone and bone implements that were far more sophisticated in their construction than anything that had been previously found. Given that the remains so far attributed to *Homo sapiens* found elsewhere in the world had been dated to be much later than these African modern humans (roughly 40,000 years ago in Europe, and 60,000 years ago in S. Asia and Australia), the conclusion seemed fairly straightforward, modern humans, *Homo sapiens*, had evolved first in Africa, and as early as around 200,000 years ago.

This remarkable discovery gave rise to what is now known as the *Out of Africa hypothesis*, a theory strongly advocated by a Professor Chris Stringer of the Natural History Museum in London, and one that is now generally accepted by most (but as you would expect, not all) trying to piece together the story of human evolution, and one that sees our recent ancestors emerging in Africa and migrating across the globe in a second great migratory wave.

The first wave, as we have seen, sees our friend *Homo erectus* leave Africa some 2 million years ago and colonize the far reaches of the globe producing Peking and Java man in Asia, and slowly evolving into *H. neanderthalensis* via a side order of *H. heidelbergensis* stock in Europe, a view that up to this point is not too dissimilar to the *Multiregional hypothesis*. But then from that point on the theory takes a somewhat different and unexpected plot twist. Scientists now believe that around 150,000 years ago a second major wave of Africans, Mr. Kibish and his friends, also got itchy feet and decided they too wanted to see more of the world, and so packed their passports and whatever warm clothes they had, and left their African evolutionary birthplace and radiated out across the globe.

Thus, by implication, the *hominids* that formed this second great wave of migration, these first African *Homo sapiens*, must then be the true ancestors of each and every one of us alive today. Based upon this *Out of Africa hypothesis* it seems that for all modern humans Africa was the "Garden of Eden", and our true ancestral homeland.

Then in 1987 some "outside of the box" pioneering work carried out by Allan Wilson and Rebecca Cann from the University of California at Berkeley provided some very strong, and much needed, supporting evidence in favour of the *Out of Africa hypothesis,* and another nail in the coffin for any remaining die-hard *Multi-regionalists*.

Now Wilson and his fellow researchers were not palaeontologists, anthropologists, or even fossil-hunters, they were in fact a whole new breed of boffin altogether, they were geneticists, and they were particularly interested in one very special type of genetic material known as mitochondrial DNA. Mitochondrial DNA is found only in those very interesting little components of nearly all cells known as mitochondria which we met earlier in our story when they were busy becoming the new and improved way for cells to use oxygen as a means to provide their much needed fuel. Now what these Berkeley DNA-brainiacs discovered was that this particular type of DNA would also likely prove very useful in the study of human evolution, and for two very good reasons.

Firstly, just about all mitochondrial DNA is passed on only through the female line, and so it doesn't get diluted with any "mud-watering" DNA from the male line and so provides something of a unique DNA history of the world's women. Secondly, it mutates about 20 times faster than the normal DNA (our genetic code) found inside the nuclei of our cells, and so making it easier to trace genetic patterns across shorter periods of evolutionary time.

What they surmised was that by comparing the mitochondrial DNA of today's modern women across various ethnic groups, swiftly followed by some clever work on the chalkboard, they should be able to work out a genetic history of modern (female) humans, a genetic family tree which would enable them to estimate when various ethnic groups split from common human ancestors, and thus hopefully trace back to the root of the *Homo sapiens* family tree (on the female side at least). So off they trotted and took mitochondrial DNA samples from women across various ethnic groups, did their analysis, and crunched the numbers, and what they eventually concluded was to be as dramatic as it was surprising.

What Wilson and his Berkeley team announced to the world was that the results of their analysis clearly demonstrated that every human on planet Earth, you, me, your next-door neighbor, Adolf Hitler, the Pope, and even your mother-in-law, every single person alive today can all trace their ancestral line back to one single woman who lived in Africa some 200,000 years ago. This "African Eve" as she was quickly dubbed by the ever enthusiastic press, was the "mother of humanity", the root of modern man's family tree, and from rough estimates the $10,000^{th}$ great-grandmother of us all.

Of course "African Eve" was merely representative of the mitochondrial root of *Homo sapiens*, but, to the joy of supporters of the *Out of Africa* hypothesis, the Berkeley findings demonstrated she very clearly originated in Africa and around 200,000 years ago, exactly the place and timeline that they had been stating for the emergence of *H. sapiens* all along. It all seemed too good to be true, and course, not unsurprisingly, it was.

Once the initial euphoria of their results had died down the ever suspicious scientific world naturally were keen to carefully review their methods, and unfortunately (and rather embarrassingly) they proved to be just a little flawed. Primary amongst these "little flaws" was that rather surprisingly the females used in the study as the sample "Africans" were not African's at all, they were African-Americans, and thus carried at least a few hundred years' worth of "Americanisation" within their genetic samples. So the suitably embarrassed Berkeley team presumably held a red-faced crisis meeting, pointed some accusing fingers at each other for a while, vowed to pay a bit more attention to what they were doing, went off and found some somewhat more "authentic" African females, and then repeated their analysis.

Fortunately their second attempt, showed very similar results to their first and so saved their high-brow blushes, and the "Out of Africa" supporters all breathed a collective sigh of relief. The Berkeley findings were then offered further support in 1994 when Luca Cavalli-Sforza, a population geneticist from Stamford University, published his book *The History and Geography of Human Genes* in which he detailed the results of his genetic study of over 7,000 human population types, and which also seemed to indicate that *Homo sapiens* had indeed originated in Africa.

Then in 1997 a team led by Svante Paabo at the Munich University somehow managed to extract some DNA from a Neanderthal bone specimen, it was only a tiny fraction of the full DNA sequence but it was enough to compare to the equivalent DNA sequence in our own *Homo sapiens* DNA. The results "strongly" indicated that there was no genetic link between modern humans and the *Neanderthals*, and so seemed proof enough that *Neanderthals* did not go on to evolve into modern man.

Interestingly, Paabo's findings seemingly gave proof that there was no widespread inter-breeding between the *Neanderthals* and any newly arrived exotic and alluring travellers from Africa, although it's hard to believe that there was not at least a few ill-advised alcohol-fuelled trysts, and some wild Stone-age parties that led to some next-morning blushes at some point. Genetic science was seemingly shooting hole after hole into the *Multiregional hypothesis,* and so it was all looking a bit cut and dried for "African Eve" and her travelling offspring as the single root of modern man.

Unfortunately, once again things are never as straightforward as they first seem, and today's ever improving techniques for genetic study are now starting to show that our genetic history is far more complex than we had first hoped, with strange genetic anomalies and variants now starting to be identified in certain small population groups that just can't be traced back to any African origin. Such anomalies continue to provide a degree of head-scratching amongst population geneticists and anthropologists alike, but seem to indicate that at least a degree of *Multi-regionalism* must have taken place, either by some regionally specific evolutionary adaptations or even maybe that some small-scale localized "hanky-panky" (willingly or not) did indeed take place with other early *hominids*, such as Neanderthals or Java man, who certainly would have co-existed for a period of time with the newly arrived *Homo sapiens*.

There are even a few bearded radicals on the fringes of genetic research who have rocked the evolutionary boat further by going so far as to claim they have their own research which supports a theory in which nearly all European humans alive today do indeed have somewhere between 1% to 4 % of their DNA derived from *Neanderthal* DNA, and so although most now agree that the *Neanderthals* did not just morph into a local European branch of *Homo sapiens*, clearly there may well have been at least a few *Neanderthals* attractive enough to those new African immigrants to warrant something more than just a polite conversation over a shared meal of mammoth steak.

However, the general consensus amongst human geneticists today is that the genetic record still primarily supports an *Out of Africa hypothesis*, but that to some lesser degree a level *multi-regional* evolution must also be thrown into the increasingly complex mix. It seems the reality is that the birthplace and eventual migration of modern man, *Homo sapiens*, is not just a straightforward choice between two seemingly conflicting hypothesis, it now seems far more likely that the truth lies somewhere between the two.

And so it seems anatomically modern humans did indeed evolve somewhere in Africa around 200,000 years ago, some of whom then marched out of their homeland in a second great migratory wave, seemingly making land in the Near East around 125,000 years ago, on to Asia round 50,000 years ago, and stopping only briefly to draw breath moving quickly onto Australia about 10,000 years later.

Europe saw its first anatomically modern human visitor turn-up at EU Passport Control around 43,000 years ago, while the Americas had to wait for some willing volunteers to make their way to Siberia then across the Bering land bridge into Alaska around 15,000 years ago before they were first graced with their presence (you have to imagine that at the time this was not likely a mission that caused too much of a crush at the Siberian visa office in downtown Southern Africa).

And from this point on it was just a few thousand years via a Bronze then Iron Age, a handful of Egyptian and Chinese dynasties, a Greek then a Roman empire, some best forgotten years in the Dark ages, followed by a remarkably fruitful Renaissance period, an agricultural and industrial revolution, a few pointless and devastating World wars, a quick stroll on the moon, and evolution rather conveniently arrives at you and me alive here today in the 21st century.

18

And Then There Was One

But we still haven't answered our question around just what did happen to the *Neanderthals*? Did those first modern humans newly arrived from Africa take an instant dislike to the shorter, stockier, hairier, more robust versions of themselves that they encountered there and so embarked on a course of "ethnic cleansing", hunting them to extinction in the same way their later descendants would treat the dodo? Or did they squeeze them out of their hunting grounds until they were literally forced over the cliff's edge of Europe into the sea never to be seen again?

Well to-date there have been no sites found indicating any mass slaughter of *Neanderthals*, no "killing fields" have been unearthed, no evidence of any Stone-age battle-fields, or any evidence at all to indicate that any *Neanderthals* ended up skewered on the pointed end of a stone-age weapon. In fact evidence suggests that having arrived in Europe around 43,000 years ago, the newly arrived *Homo sapiens* lived alongside their *Neanderthal* evolutionary cousins for at least 10,000 years before those cousins checkout of Europe and the evolutionary map altogether, and so all-out conflict would not appear to be the cause of their ultimate extinction.

We must remember that 40,000 years ago Europe was not populated with hundreds of millions of people (Neanderthal or African tourist) in the way it is today. It was a sparsely populated ice-age tundra at best, and encounters between Neanderthal and new "Modern man" were likely infrequent and fleeting. No, to understand what happened to the *Neanderthals* we must first understand a little more about those early modern humans who arrived in Europe fresh from the savannahs of Africa.

Early European Modern humans (or EEMH's as the boffins like to call them) were first discovered in 1868 in a rock shelter in Southwestern France named Abri de Cro-Magnon, and the term "Cro-Magnon" soon became generally associated with these first known modern people in Europe. Many similar specimens of *Cro-Magnons* have subsequently been found across Europe as a whole with the oldest dated to be around 43,000 years old.

Physically these new Europeans were remarkably similar to you and I, they were straight limbed and tall compared to the indigenous Neanderthals, with the anatomically modern facial features of a straight forehead, only slight brow-ridges, and a prominent chin. Their brain capacity was actually on average slightly larger than today's modern humans at around 1,600 cc, and they were likely tan-skinned given their recent heritage. The only real difference we would notice between the *Cro-Magnons* and what you and I see in the mirror today (apart from the high-street fashion) is that they were rather more robustly built and physically more powerful, although this should be of little surprise given their stone-age hunter-gatherer way of life compared to our 21st century lifestyle which generally means the most exercise many of us get on a daily basis is to lift the TV remote.

But it turns out the *Cro-Magnons* were not going to be just physically different from the *Neanderthals*, initially however, around 40,000 years ago as the *Cro-Magnons* were just starting to settle into their new European home, there seems to have been little to choose intellectually between the creativity, the tools, and the general way of life of the *Neanderthals* compared to those early *Homo sapiens*. But then almost overnight it seems a cultural and intellectual switch was flipped in these new African immigrants.

In the space of just a few thousand years *Cro-Magnon* tools appear to become far more sophisticated, more intricate, more specialized in their construction and use, cave paintings appear at sites such as the Grotte Chauvet and Lascaux which show an art-form now rich in sensitivity and perception, painted with perspective, shading and relief, and showing a real artistic awakening that was also manifest in intricately carved sculptures of wood and ivory and jewellery in the form of ornate necklaces. It appears that almost overnight these early humans had discovered art and culture, and were now expressing themselves in far more sophisticated ways than had previously been seen by any *hominid* up to that point (including Mr. Neanderthal).

And this "awakening" was not confined to just an ability to express themselves through art, *Cro-Magnons* now started to evolve far greater and wider social interactions, communicating between social groups spread across hundreds of miles, sharing ideas, tools, thoughts and knowledge, transporting materials from sites and trading across huge distances, and thus creating social bonds and societies across something wider than just their immediate dependants, in short *Homo sapiens* began to network.

But what happened to cause this intellectual and cultural surge? Certainly when *Homo sapiens* first set off from their African homeland to conquer the world there was no indication of any such "awakening". Scientific bods guess at perhaps some incredibly fortuitous genetic transformation in brain function kick-starting such an intellectual explosion, but we might just as well put it down to some "divine intervention", Alien genetic tinkering, or even the sudden appearance amongst the ranks of *Homo sapiens* of a God-like messiah who showed them all the way to self-awareness and intellectual superiority. The reality is we will likely never know how a species with an intellectual capacity on a par with that of the *Neanderthals* when they left Africa (being roughly that of an average 5th grade student) and yet arrived in the four corners of the globe with their ranks now swelled with the equivalent of Leonardo Da Vinci, Albert Einstein, and Abraham Lincoln.

But whatever the catalyst for such change it is clear that sometime around 40,000 years ago modern man woke-up one sunny morning having experienced a creative vision that left the *Neanderthals* in their wake. While the *Cro-Magnons* were busy discovering their consciousness, communicating and trading across vast distances, building a broader society built with social structure and collective benefit, and expressing themselves in more sophisticated and intricate ways through their art and tools, the poor *Neanderthals* it seems simply chose to continue to operate in small close-knit isolated bands, living in narrow ranges of around 25 miles within which they looked to source all their needs, and communicating little across other neighboring bands.

And so in Europe it seems that the *Cro-Magnons* gradually took a grip over their environment, passing on information around the best hunting grounds, the best quarries for sourcing materials for tools, trading across vast distances, building social networks and structures for collective mutual benefit. The *Neanderthals* meanwhile remained somewhat isolated choosing to continue to exist in far smaller individual and somewhat independent pockets of populations which were forced to deal with not only the challenges of ice-age Europe but now the influx of the well-organized intellectually superior machine that was the *Cro-Magnons,* who were now competing with them for the same European Real Estate.

Cro-Magnon populations duly exploded, and likely within the space of a just few thousand years vastly outnumbered *the Neanderthals,* and thus very quickly just out-competed them by both sheer weight of numbers and intellectual superiority. Squeezed into ever more isolated and difficult parts of Europe the *Neanderthals* it seems died-out not so much in any blaze of glory "Butch and Sundance" moment but likely with a slow whimper, and within around 10,000 years of the *Cro-Magnons* first arriving in Europe the *Neanderthals* had been squeezed out of Europe and consigned to the history books.

The *Neanderthals* had survived in Europe for well over 200,000 years, they were tough, strong, physically well adapted, and damn near indestructible, but their evolutionary response to dealing with the challenges of ice-age Europe, one which had stood them in good stead for nearly 200,000 years, was finally undone by the influx of a species intent on not just surviving but also evolving intellectually, spiritually, technically, one which looked to evolve societies built around communication, trade, and social structure. A species that was not just concerned with its needs to survive today but one which was also planning, growing, and developing for the future, a species thinking about their place in the world and expressing those thoughts through art and religion, a species that had awakened the human soul. In reality, as soon as the first *Cro-Magnons* set foot in Europe there was only ever going to be one outcome, and from that point on the poor *Neanderthals* were living on borrowed time.

And so roughly 30,000 years ago *Homo sapiens* suddenly found themselves as the only known human species left drawing breath on planet Earth, the first time in our 7 million year human evolutionary history that any *hominid* species had not had to share space on our planet with other *hominid* species.

As modern humans we are used to the idea of being the only "humans" on the planet, and thus we see ourselves as somehow special, unique, and somewhat apart from all the multitude of other species to be found inhabiting planet Earth. But since splitting from our chimpanzee cousins to kick-start our own evolutionary line we have spent more than 99.5% of our existence sharing our planet with at least one, and usually several, other representative species from the numerous other branches in our evolutionary tree.

The last few chapters have attempted to trace (admittedly at a very simple level) that complex evolutionary tree from those first *hominids* that split from our common ancestor shared with the chimpanzees some 7 million years ago, through Ardi and his first nervous steps onto the savannahs of Africa, on to Lucy and the Turkana boy, through to those first anatomically modern humans that set out from their African birthplace to conquer the world, and in the process consigning all other living branches of our human family tree to the evolutionary waste-bin.

It's was an evolution driven by the need for continual adaptation to climatic and environmental change, and one that we have seen unfold in 3 distinct phases. Firstly through the ape-like *Australopithecines* who evolved to become the first true bipeds, through the tool-making of *Homo erectus*, and finally the arrival of modern humans whose intellect, creativity, and social bonding, ensured they would survive to populate the four corners of the globe, and become our planet's intellectual masters and sole remaining *hominid*.

In reality our evolutionary path has been a complex one that is by no means fully understood even today, and one which paleoanthropologists, and more recently geneticists, are still slowly helping to piece together, but as we have seen, each new fossil find or DNA analysis can often muddy the waters more rather than clarify our understanding. What is clear however is that the story of human evolution is far from being a tale which draws a simple evolutionary straight line from some tree-dwelling ape to you and me, it is more akin to a 5,000 piece jigsaw puzzle of a white cat in the snow, a jigsaw where so far we have managed to put together about half the pieces but unfortunately we have no idea where the remaining pieces are even to be found.

Although we do have some broad agreement as to the "likely" path from ape, to ape-man, to man, amongst those who dedicate their lives to trying to piece together that jigsaw puzzle, it all does highlight the quite remarkable fact that we actually know less about the evolution of our own species over the course of a few million years than we do about the Earth's own 4.5 billion year history.

19

Burnt Fingers and Singed Eyebrows

It took our ancestors several million years to start thinking that maybe life on the ground might just offer better opportunities for long term survival than life in the trees, leave our chimp cousins behind, step down onto the savannah, stand up, and finally commit to a life on two legs, and even longer to learn to communicate with each other with something other than one grunt for "Yes" and two grunts for "No". But once evolution had finally got us on the ground, upright, and chatting away, progress happened pretty quickly.

Effectively from that point on we decided not to wait around in the hope that evolution would continue to favour us, instead we started to take matters into our own hands (quite literally). Rather than wait around to see if evolution would happen upon a chance genetic mutation in the next million years or so that meant we might develop legs or arms long enough for us to reach up to pick the fruit from the top of the tree, or a stomach that could safely digest two week old meat without the fear of having to see it all again a few hours later, or even evolve a skin that was waterproof, we just went ahead and invented the ladder, the fridge, and the umbrella. Thus, within the space of just a few hundred thousand years we were painting the walls of caves, creating ornamental jewellery out of shells and bone, and eating three square meals a day from the food we had just killed, sliced, diced, and cooked with our new state-of-the-art stone tools.

Within the space of another few short millennia we were mastering the art of farming, forging bronze and iron, building our own shelters, living in communities, and fashioning clothes beyond just simply throwing on the hides and fur of the animals we had just recently killed and then eaten.

And from there it is was just a short intellectual hop to the pyramids, the glories of Athens and Rome, ships, telescopes, and then quickly on to electricity, radio, television, cars, planes, radar, space rockets, and the wonders of sliced bread.

There have of course been thousands of other inventions and discoveries that have made a significant difference to mankind, our ability to adapt and master our environment, enhance our prospects of survival, further our understanding of the world we live in, and generally make life far more agreeable for you and I. Take the printing press for example, the invention of a German Blacksmith in the early 1400's, which was responsible for the information and knowledge revolution and which kick-started the progress of the Renaissance. Or the double whammy by the Han Dynasty of 2nd century China, inventing both the compass and paper, with which the world could now be far more easily explored, and scholars could convey their ideas in convenient handy sized fountains of wisdom which the more intelligent amongst us will know as "books".

And then of course there was the invention of numbers, without which there would be no way to count, and with no way to count there would be no math, with no math there would be no engineering or science, and without which there would be nothing to keep our brightest bods busy, and thus no foundation on which to build most of the advances and inventions of modern civilization. Now it's also true to say that without numbers there would be no money, and thus no banks, and thus no overdrafts, and thus no debt, and thus no greed, and thus a blissfully happy world, maybe.

However, the reality is that the invention of numbers, and thus counting, and then thru to mathematics in all its current bewildering and mind-boggling incarnations, is the cornerstone of much of the scientific and technological progress that came after, and thus clearly one of the most important steps mankind has taken. Numbers provided the basis for trade, architecture, engineering, our observations around time and scientific experiment, how we organize our lives, and how we bring order to our world. Without numbers and mathematics the likes of Isaac Newton and Albert Einstein would have been left with no way to prove their theories, children would have no way to know that the candy has been distributed fairly, and parents no way to know how many children they needed to account for each night at bedtime. The use of numbers permeates every aspect of our lives.

Let's also not forget the wheel, an invention we do tend to think of as likely being pretty high on our early ancestors "things we really need to invent" list (probably just ahead of an urgent need for soap forced by the close proximity of mankind's new vogue for communal living). But surprisingly, several significant inventions likely pre-dated the wheel by thousands of years such as sewing needles, woven cloth, rope, basket weaving, even boats, although until the wheel was finally invented boats were presumably limited to a size that could easily be pushed, pulled, or carried to the water's edge.

Ask most people and their perception of who, how, and when, the wheel was invented revolves (no pun intended) around a vision of a Stone-age man about two million years or so ago, sitting in a cave high on a hill, chiselling away at a huge flat stone rock fashioning it into a round shape which he then stood on its side (presumably to admire his handy-work), and accidentally letting it roll down the hill, and thus inventing the wheel.

In reality, the earliest examples of the wheel so far discovered by archaeologists appear as late as the Bronze-age around 3500 BC in Mesopotamia, near the Black sea, and although the wheel may not have arrived on the scene until relatively late in our human history, once with us it changed man's future forever. If we look around today we see use of the wheel everywhere and not just in the obvious transportation examples of cars, bicycles, and wheel-barrows, but also in the mechanisms for gears, watches, steam turbines, and flywheels. The wheel is the basic principle behind almost every mechanical device in use today, indeed, it is the power of the wheel that drove the great Industrial revolutions of the 19th century.

Or consider the power and impact of the computer in our daily lives, a machine which finally freed scientists from their blackboards, and secretaries from their typewriters, and created a whole new way for kids to waste away their free time, an invention itself that has evolved since the 1940's from a machine the size of a small building and consuming the power of a small town, that can now fit in your pocket powered by a small battery.

Now ask your average spotty-faced teenager and they will likely cite the Nike running shoe, the iPod, the mobile phone, and the skateboard, as the pinnacle of man's achievement, some of our more mature citizens may more realistically point to electricity, radar, or penicillin, and those with a more ascetic leaning may even cite the Electric guitar, Concorde, or even the bikini. But clearly as important as these all are there are but a handful of inventions and assorted eureka moments that we can truly say changed the course of mankind's destiny (although most red-blooded males would still likely try and make a case for the bikini).

These are the giant leaps forward that not only by their own virtue changed mankind's ability to survive and master the world around us, but also as a consequence also paved the way for much of mankind's subsequent advances. Without these few key discoveries and inventions we would likely still be living in caves, living off nuts and berries, still wearing the latest line in mammoth skin coats, and wondering why we ever got down from the trees in the first place. And right up there in everyone's top ten list of "the most important things ever invented", or at least in the lists of anyone who doesn't ride a skateboard to work or play an electric guitar for a living, is the stone-tool.

Up until about 2.5 million years ago, our early ancestors survived off a rather uninspiring daily diet of nuts, berries, and grubs, but it seems that around this time those same early ancestors also started to develop a taste for something a little more filling, something a little more protein rich to feed their now developing brains, what they wanted was meat. But the problem was that for those meat-craving early ancestors to eat something other than the "tree-hugging friendly" nuts, berries, and grubs, they first needed to kill the current owner of the required fat and muscle, which is clearly not something they were going to let happen without something of a fight.

But to hunt and kill a sizable enough animal to make the whole effort worthwhile, the "law of the jungle" requires you to generally have the ability to outrun your prey, and also possess very large, very sharp teeth and claws to kill them with once you had finally caught them, and our early ancestors unfortunately had neither the ability nor the attributes. Thus their only other alternative was to hope that they stumbled across a fresh supply of meat conveniently left by another "not quite as hungry as they thought" carnivore clearly more adept at hunting, or resign themselves to yet another day without meat, find the nearest convenient tree to hug, and settle down for some more nuts and berries.

Thus, those early ancestors were forced to exist primarily as scavengers given that their physical attributes were clearly a poor match for both any prey and their fellow meat eating competitors. And so unfortunately, it seems our early ancestors were a little further down the food chain pecking order than their evolutionary ambitions would need them to be. What they needed was a competitive edge, a means to tip the scales in their favour so they could depend less on scavenging and other predator's scraps, and become the true hunter gatherer that evolution demanded of them. The solution it turned out was just around the corner, well more accurately lying on the ground right in front of them, the solution was the use of rocks and pebbles to build stone-tools.

With the use of such tools mankind suddenly had a competitive edge, and he could suddenly accomplish tasks that the human body on its own could not, such as deliver a killing blow to a prey, scrape and cut the hide and meat from that killed prey, and chop wood for fuel and construction. Clearly it is much easier to kill a 200lb antelope with stone spearheads and axes than it is to attempt such a feat with just raw determination and your bare hands.

Now we should first note that many animals today also use tools to access food or for grooming, and we are not talking about those circus elephants or sea-lions that perform tricks with hoops and balls learned to gain reward, but animals in the wild demonstrating a level of intelligence to gain some advantage by using a tool such as a stick or a rock.

Chimpanzees have been observed using sticks as probes to collect ants and to break bee-hives to extract honey. Gorillas have been seen using sticks to measure the depth of the water they are wading thru, and monkeys are known to use stone hammers to crack nuts. Even sea otters are known to use stones to hammer abalone shells on rocks, and some enterprising birds have been observed dropping oyster shells onto hard surfaces to crack them open, with some even risking life and the threat of a jaywalking ticket to place nuts on roads so passing cars will run over them to release the bounty inside. However, although clearly demonstrating the use of primitive tools what these smart members of our animal kingdom are using by way of those tools and how they go about using them does not compare to even the most primitive implements that show up in our early ancestor's history about 2.5 million years ago.

Part of their challenge is that the hands, feet, claws, or flippers, of our animal friends are nowhere near as dextrous as that of the human hand, and as such make fine motor skills difficult (if not impossible). Our highly adapted hands which allow modern *Homo sapiens* to throw a curve ball or play a Chopin Nocturne, are in a large part the key to our ancestors' ability to fashion stone-tools beyond the most basic tools used by today's other earthly creatures. Fortunately it seems we are a long way from today's apes progressing much beyond an ability to use sticks and rocks to help crack nuts, and thus thankfully then under little threat of a "Planet of the Apes" scenario any time soon.

But the difference is also reflected in some more subtle, but equally important, advances in thinking that also seem beyond the abilities of even Russia's early astronaut monkeys. The makers of even our most rudimentary stone-tools appear to have demonstrated an ability to first choose the most appropriate rocks to use, and to also understand where and at what angle to strike the stone to produce the best results. It seems not only did we have a hand well adapted for such work, but we also began to put some thought into the process before starting. Each tool was not the simple fortunate consequence of multiple "random" attempts, but the result of a well-crafted process implemented by a superior physical adaptation.

Precisely when early humans started to use stone-tools is difficult to determine, because the more primitive these tools are (for example, naturally sharp-edged stones just found on the ground) the more difficult it is to decide whether they are just natural objects or tools fashioned by man. But apparently our archaeologist friends seem to be somewhat in agreement that the first discernible stone-tools turn up around 2.6 million years ago, when it seems that our early ancestor, *Homo habilis,* used tools made out of round pebbles that had been split by simple strikes to form the required sharp edge. These very crude stone-tool from this period are known as Oldowan tools, named after the site in the Olduvai Gorge in Tanzania, where they were first discovered in the early 1930's.

In contrast to those early crude Oldowan tools, mankind soon developed a method which allowed the tool-maker to have far greater control over the resulting tool. These new high-tech tools were worked symmetrically and on both sides indicating far greater care in the production of the final tool. Such tools are typically found with *Homo erectus* remains (the most advanced predecessor of modern humans of the time, and thus the most likely architects of their development), and were the dominant tool-making technology for the vast majority of human history, with the earliest accepted examples found in Kenya dated to around 1.7 million years ago.

With their distinctive oval and pear shape, these tools are known as Acheulean tools after the area of Saint-Acheul, a suburb of Amiens in France, where such tools were first identified in 1859, and had attained a very high level of sophistication, required a lengthy patient process (although the Stone-age did likely provide ample spare time for the would-be tool maker to pursue such craft), had a much larger cutting edge than their Oldowan precursors, and were used more for hacking, slicing, cutting and scraping, rather than used in any heavy crude percussion blows which likely would destroy the edge.

As confirmed by clever men with beards, over the course of the following million or so years there was a rapid succession of increasingly more complex stone-tool technologies, such that by around 10,000 years ago stone-tool making had advanced to the manufacture of what people who are passionate about stones call microlith tools. These are smaller stone tools crafted to form points, and which were fastened to wood or bone shafts with resin or fibre to form composite weapons or tools such as spears, arrow, or harpoons. Single microliths were now used to fashion arrows and spears, with multiple microliths attached to spears and harpoons like teeth to create composite tools, resulting in far more effective hunting (clearly it's far less dangerous to throw a spear at your prey from the safety of some distance than it is to have to get up close enough to hit it) and thus this period in mankind's history witnessed a change to far more intensive hunting and fishing with an associated increase in social activity, and the development of more and more complex settlements.

Mankind had very quickly become completely reliant on stone-tools for his own survival, and almost overnight they elevated him to the very top of both the food chain and the evolutionary tree. Stone-tools provided him the means to fashion hides for warmth, chop wood for fires, and to hunt for food. It's likely then that without the invention of stone-tools mankind as a species may well have ceased to exist long before he'd even got out of second gear, which would have been rather good news for the numerous other species we later went on to hunt to extinction with those same stone- tools, but not so much so for you and me.

However, of all mankind's key discoveries and inventions, probably the single most important of those was our discovery of how to master and control fire. Now, despite the belief of many, man did not "invent" fire, and neither did anything, or anyone, else for that matter. One of our early ancestors did not wake up one cold stone-age morning shivering under his mammoth-skin blanket, proclaimed through chattering teeth that "enough was enough" and determined to devise something new that would keep him and his family somewhat warmer over the long winter months, there was no "Eureka" fire moment for mankind. No, fire had been around as a natural phenomenon since the Earth first coalesced from the gas and dust resulting from the creation of the Sun some 4.6 billion years ago, long before our ancestors took their first nervous steps onto the plains of Africa, and certainly long before they first burnt their fingers trying to discover just what this strange source of heat and light was.

Actually, fire itself doesn't exist at all in a natural form, it isn't matter at all, you can't just pop down to the local shops and buy some fire. What fire is, is a visible, tangible side effect of matter actually changing form, it's a chemical reaction. Fire results from a reaction between the oxygen in the atmosphere and some sort of fuel such as wood or kindling. Of course, wood doesn't just spontaneously catch on fire just because it's surrounded by oxygen (that would prove a little bit of problem for the forests of the world given our atmosphere laden as it is with 21% oxygen), no for the reaction to happen, you have to heat the fuel (such as wood) to its "ignition temperature", and thus something first needs to heat the fuel to this temperature. The heat can come from lots of different things such as a match, friction, lightning, or something else that is already burning, and when, in the case of wood, it reaches about 500 degrees Fahrenheit, it reacts with the oxygen in the air, miraculously ignites, and thus burns. The side effect of this burning is a lot of heat and light, and as those first inquisitive humans quickly discovered, a great deal of pain for anyone who gets too near to its source.

And so the occurrence of fire itself happens quite naturally in our world as a result of such phenomenon as volcanic activity, meteorites, or lightning strikes which coax the fuel source to its required ignition temperature, and thus it needs no intervention from man, beast, or arsonist, fire will occur regardless of whether we are present at the event or not.

Indeed most land animals are aware of fire and adapt their behavior to it, admittedly this behaviour usually takes the form of running as far away from it as possible, but it still demonstrates recognition of fire as a natural phenomenon and its impact on anything that comes close to it. Thus, our ancestors would have been aware of fire long before they could make fire on demand themselves, and thus would have quickly come to learn of its benefits in respect of the warmth and light it gave. But it would not have been until early man discovered the ability to handle, manage and control fire himself, would he have been able to fully appreciate and gain evolutionary benefit from it.

The control of fire by early humans was a monumental turning point in human evolution that allowed humans to cook food and obtain much needed warmth. Making fire also allowed the expansion of human activity into the dark and colder hours of the night, and provided protection from predators who to this day continue to treat it as something to be avoided at all costs. By controlling fire mankind overnight elevated himself straight to the top of the evolutionary tree, and left all of Earth's other creatures, till that point still equal participants in the daily fight for survival of the fittest, in our dust (or more accurately, ash).

And it wasn't just the obvious and immediate advantages of suddenly being able to provide light and heat on demand, the instant protection it offered, or the opportunity to cook the latest mammoth kill into something a little more palatable and something that didn't require 3 hours of constant chewing to swallow. No, it was what the control of fire also meant for mankind's future. It was through our ability to handle fire that all future technological advances were made possible, as through the control of fire eventually came an understanding and practical application of metallurgy, the ability to fashion bronze and iron tools, and thus to the dawn of science.

It's likely that early attempts to control and manage fire resulted in numerous burnt fingers and singed eyebrows, but clearly our early ancestors persevered, and there is evidence for the controlled use of fire by our early cousins *Homo erectus* beginning some 400,000 years ago, while claims regarding earlier evidence are mostly dismissed as inconclusive or sketchy. However evidence of controlled fire becomes widespread around 50 to 100 thousand years ago, suggesting widespread and regular use from this time, with our ancestors all clearly now safer, warmer, and sharing campfire stories late into the night.

Now the first and easiest way to control fire would have been to use the hot ashes or burning wood from an existing forest or grass fire, and then to keep the fire going for as long as possible by adding more wood and plant materials throughout the day, but, what man really needed was a way to start a fire from scratch, independently. Now given that matches and the Zippo lighter would not be invented for at least another few hundred thousand years or so, our early ancestors needed to come up with an alternative method to light their fires while they awaited for such fire-starting convenience to arrive.

The boffins of course can't be sure, but likely the first methods involved the use of friction to generate the heat to ignite dry tinder. Fire can be created through friction by rapidly grinding pieces of wood against each other or a hard surface. Any type of wood can be used, though softer woods work better, a method still taught today to boy Scouts the world over. The flint method, where hot sparks are struck from a piece of flint onto suitable tinder and fanned into flames, was also likely used by primitive cultures. These methods have been known since the Stone-age, and are still in common use today by some indigenous peoples in remote corners of the globe, people who someone had clearly forgot to tell that the long wait for matches and the lighter to be invented was now over, and that they were available at a bargain price down at their local trading post.

As time marched on controlling fire lead to an understanding of basic metallurgy and thus propelled mankind from the Stone-age thru the Bronze and Iron ages with all their advances in tools, machinery, and associated understanding of elements and the sciences, that all ultimately led too much of the technological advances of today. In addition, fire also formed the basis for communal living, initially driven by a need for collective advantage, that itself then led to the agricultural revolution and ultimately becoming the catalyst for today's modern societies.

Early man's decisions to step down from the trees and stand upright on two legs was arguably his single most important "decision" in his natural evolution to today's *Homo sapiens*. But of all the technological advances and inventions attributable to mankind's progress from upright-ape to Mission Commander on Apollo space flights, it is the control of fire that was very likely the single biggest catalyst for human advancement. We will never know who was that very first individual to discover how to control fire, we will never know his (or her) name, what they looked like, or indeed how much hair they singed in the process, but very likely they are the single most important individual to have lived throughout our five million year history.

20

Homo Perfectus

Let's take a little leap of faith for a moment and assume that our own little species was indeed created under some glorious divine providence driven by an all-powerful celestial leader. A leader who, to protect the innocent, we will call Zod.

Now let's further assume that one day the mighty Zod comes to you and hands you a blank piece of paper and asks you to come up with some ideas to help redesign the current *Homo sapiens* model he has got out there in the world. He thinks there may well be some room for upgrades and is looking for ideas around a possible leaner, meaner, *Homo sapiens* Mark II. Now to be fair to Zod, we are given to believe that he did only have a week in which to create the Universe and everything in it from scratch, and given that he was likely rushing so that he could also take the seventh day off, (presumably Mrs. Zod needed some errands running), he probably did rather rush to finish the Mark I model so as to get us of the production line without the need for overtime.

Now remember that we are also told that *Homo Sapiens* Mark I was also made by Zod in his own image, so you can't stray too far from the current look and feel (you don't want to piss-off the boss, particularly when that boss is Zod himself) so your design for the Mark II model needs to stay fairly true to well, how you look now. However, Zod clearly feels there may be some room for potential improvements, so you thank him for the opportunity, take the blank sheet of paper and retire to a quiet corner to scratch your head a little and have a think. So what, if anything, would you change in our current physical form to improve upon it, to make it better suited for life on 21st century planet Earth?

Would you look to improve upon our senses? A more dog-like nose perhaps, or additional eyes for a more 360 degree panoramic view of the world, or adopt more elephant-like ears or a large lizard-like tongue for improved hearing and taste? Or would you look to be a little more radical and propose new more advanced features such as sonar echo location, or add two extra legs that remained conveniently tucked away up in your torso until needed, at which time they would drop down like aircraft under-carriage to ensure we could all run as fast as Usain Bolt? Or maybe look to be less radical and just look for some additional fingers and toes. Would we be better taller, shorter, wider, thinner, or with a brain enlarged to the size of a watermelon? Or, do we just need more big-breasted blondes with swimsuit bodies, and fewer people with an inexplicable need to wear socks with their sandals?

Now regardless of whether Zod actually exists or not it's fairly unlikely that He (or anyone like him) will be coming to you with a blank piece of paper any time soon, and as we have already seen humans, just like every other species, are merely the result of millions of years of natural selection, but if you really think about our current physical form you will, outside of the outlandish tongue-in-cheek suggestions above, likely find it hard to see where true improvements could actually be made.

Right at the start of this brief tale I referred to the six key adaptations that have been widely cited by the more intelligent beings amongst us as being crucial to the unique success of our species above all others past or present; our intelligence, our linguistic skills, our superior visual ability, our incredibly dextrous hands, our ability to move around comfortably on two legs, all nicely topped off with our innate social nature. Not a bad collection of tricks to assemble over the course of several million years or so.

Mankind's intelligence, his gift for abstract thought and inquisitive nature, have led him down a path where he has sought and found solutions to many of the challenges placed before him that otherwise may have slowed, or even placed a sizable brick wall in the path of, his evolutionary march. The cognitive abilities of our early ancestors ensured that we mastered fire, built simple tools, and over the course of millennia invented progressively more elaborate tools to overcome the environments daily challenges.

And yet for all our great ingenuity and invention, the greatest tool mankind possesses is still that incredibly manipulative tool to be found conveniently stuck on the end of each of your arms, the human hand. No other animal possesses such a wonderfully adapted appendage for intelligent exploration and manipulation of the environment in which we find ourselves. Only our ape and chimp cousins come close to such a dexterous hand, but although possessing an opposable thumb (the key to our hands success), they do not possess the same ability for fine motor movement and control. A chimp can pick up a stone and pick fleas out of their mate's hair, but they could not thread cotton through the eye of a needle.

As for standing upright on two legs, imagine for a second we instead had evolved to stand and walk on all fours, "knuckle-walkers" like our ape cousins, or the back-row of a French Rugby scrum. Such a dextrous tool as our hand would then be very limited in its use as it would rarely be free to demonstrate what a miracle of nature it actually was, they would be forced to spend most of their time occupied in helping stop us from toppling over, not ideal. Thus, our bipedal posture frees our hand to do what it does best. It is impossible to envisage mankind progressing much beyond the Stone-age without the advantage and blessing of the human hand and the ability to stand and walk upright allowing it to perform its dextrous magic.

Mankind's visual ability is no less significant. Without the incredible power of the human eye, our hands would have been far less of an evolutionary advantage. If you can't see the eye of the needle properly you can't thread the cotton through it, no matter how skillful your hand. The acute sense of sight is also a primary driver for an inquisitive nature, it is seeing all the wonder around us that stimulates the cognitive process in the first place.

Technological progress is usually the result of a single mind, or the result of building upon the efforts of several earlier singular notions culminating in one man's synthesis of the sum into a technological advance. Rarely, if ever, do we decide as a community to call a town-hall meeting or all gather around the campfire to thrash out some ideas to solve a particular problem we all collectively face. We usually leave such cognitive leaps to the boffins of the day who are good at that kind of stuff, and who enjoy nothing more than stroking their amply bearded chins and starring at chalkboards full of complex equations for years on end in an effort to solve such vexing problems. But once the bearded boffin(s) have found a practical solution to a particularly vexing daily challenge, unless he has some way to communicate his solution to the wider population, it will be of little benefit to the rest of us. Enter our power of speech.

In addition, no other species possesses anything remotely close to our highly evolved communication system (or at least any that we've so far managed to understand at any rate with the jury still out on the whales and dolphins) that enables us to not only communicate direct information from the perspectives of the past, present and future, but to articulate abstract thoughts and concepts in a meaningful way allowing us to readily share knowledge, discuss, argue, and collaborate to overcome our species' evolutionary challenges. We should note at this point however that like all our other key adaptations, human language (all several hundred variations of it) although providing us a key competitive edge doesn't in itself ensure our survival. Unfortunately, if a moon size asteroid does appear in our skies hurtling towards Earth there is probably little we could do to survive the imminent extinction event. Mankind collectively coming together to shout at it to "bugger off" in a 100 different languages is unlikely to meet with much success.

Language also presumes a highly social animal, something that we do have in common with dolphins, whales, and today's other primates. Our highly developed language skills would provide little competitive edge if we all avoided each other like the plague (although anyone who regularly travels on London's Underground system where you are avoided like a rabid dog if you try to converse with a stranger, would question such an innate social nature). Sociality also ensured that as a species we (usually) act to protect not just ourselves but our species as a whole, although the likes of Adolf Hitler, Attila the Hun, the 16th century Spanish Conquistadors, several European kings of the middle-ages, and the Roman Emperor Caligula, may well suggest otherwise.

It is in large part the collective synergy of these physical adaptations afforded our particular species, (bolstered as we have seen by a fair portion of good luck along the way), that has over the course of 3 million years and change, led us from Stone-age cave-dwellers to the high-tech, coffee swilling, modern city dwellers of today.

So could the mighty Zod actually do with a Mark II version, could 21st Century *Homo Sapiens* benefit from some further evolutionary tweaks. Would an extra finger on each hand improve our dexterity? Would X-ray vision, or a brain twice its current size improve our species chances of survival? Well actually, likely not.

The Chicago blues guitarist Theodore "Hound Dog" Taylor reportedly possessed six fingers on his left hand, yet despite his legendary guitar technique, it seemingly owed nothing to the extraneous digit which he apparently never saw the need to utilize. Superman's X-ray vision would clearly be a benefit for those of us with Super-hero aspirations, but for everyday use it would provide little benefit beyond that which technology could already provide. Albert Einstein, a man held as a shining example of human intelligence at its finest, surely possessed a brain physically superior to the rest of us who still remain baffled by his theorems and hairstyle, but it appears that a post-mortem examination of his brain rather disappointedly revealed a noggin of only average size and weight.

No, it would seem that we are actually exactly the optimum size, shape and weight (well for those who can keep away from the call of donuts at least) with the right number of limbs and the optimal number of digits stuck on the end of each, for a land-based life-form on a planet such as Earth with its unique oxygen rich atmosphere and gravitational force. Certainly, from where Charles Darwin sits (or sat), mankind does seem to have reached his evolutionary summit. We have not changed now as an evolutionary species for thousands of years. We haven't further evolved to grow a third eye recently so we can see out the back of our head, (however some primary school teachers do seem to be making huge strides in this area), or developed a new sixth sense (although some disciples of Eastern religions would claim to have done so), and with the exception of "Hound-dog Taylor, we all still have only ten fingers and ten toes.

Of course, when compared to our recent cousins from ancient Rome or Athens, we are clearly now living longer, grow taller, and run faster, but these are changes brought about mostly by technological and scientific change (better nutrition, better health care, Nike running shoes), not genetic mutation and evolution. The average Roman was just over 5 feet tall, and unless his chosen profession was that of Gladiator he lived to be about 40 years old, and generally died from either syphilis or a rather unpleasant wound inflicted by a marauding Celt probably upset that a new Roman Baths now sat where his Mother-in-law's pig sty used to be. Not something we as a species needed to specifically evolve to cater for.

And there are also clearly other examples of species evolving to a point where they now also seem almost perfectly designed for purpose, species that have not needed to evolve beyond their current genetic incarnation for thousands of years. They, it seems, are now also ideally suited to survive. Sharks for instance have been around under their current streamlined model for millions of years, perfectly adapted to kill and then eat whatever unfortunate life form has the misfortune to swim past their hungry gaze (shark's will basically eat anything with a face, they're nature's ultimate eating-machine).

Cockroaches similarly have seen many other species come and go without themselves having a need to evolve beyond their current design (even though from a practical sense somewhat unnecessary, a slight tweaking in the looks department may not be a bad evolutionary next step). Dolphins we all presume are a highly evolved species, intelligent and social (many believe evolved beyond that of ourselves), again seemingly without needing to evolve further.

Yet despite reaching this evolutionary peak, we don't see schools of sharks frantically trying to manufacture tools to enable them to catch and eat their food faster or more easily, or pods of adolescent dolphins trying to learn how to solve quadratic equations. And when Neil Armstrong first took that "Giant leap for mankind" onto the surface of the moon, he didn't see a cockroach already stood there waiting to greet him. (Or maybe he just didn't look under the right rock?)

So then why our own need as a species to constantly look to further "adapt" ourselves and particularly our environment beyond the place evolution has carved out for us in the natural order of things? After all, as a species we didn't turn out too badly just letting nature do its thing. Clearly as a physical species on land (like the shark in the ocean) we have evolved to the point where we are ideally suited to survive, we are at the top of the evolutionary tree, at least on land (I don't fancy our chances of surviving too long in the water in an ocean full of sharks). We are good, and so in evolutionary terms its job done.

Well certainly in the first instance the need to adapt was driven by a need to survive. Each of planet Earth's species are hard-wired to survive, and thus will attempt to do so by any means at their disposal. Our early ancestors were merely making best use of the unique adaptations evolution had kindly endowed upon them when they looked to master fire and fashion stone-age tools in order to survive.

However, it's also clear that we quickly moved on from merely creating the tools needed to adapt and thus better survive the lottery of nature's "survival of the fittest" operating model. We continued to seek to progress, invent, discover and thus evolve, beyond the need for mere survival (slapping up some nice paintings on the walls of a few caves, knocking out some ivory necklaces, and fashioning a decent pair of leather shoes, were never really going to turn the tide of evolution in our favour), and for an explanation as to why, we must at least in part return to our earlier mentioned key evolutionary adaptations.

By the time our early ancestors were mastering fire and fashioning stone tools for survival, we had already evolved a brain capable of superior intelligence and cognitive thought that was imparting a thirst for knowledge and gifting the ability for abstract thought, add to this our naturally inquisitive nature and the ability to closely observe the wonders world around us, and we can easily start to understand that human advances beyond that of a mere need for survival was likely just a natural consequence of who we were rather than any conscious choice.

We had effectively evolved into a species that was now "hard-wired" not only to survive, but also hard-wired to be inquisitive, to question, to need to understand and explain the world around us. The natural consequence of 3 million years of such "hard-wired" behaviour has led us to the myriad of mankind's inventions, the Sciences, the Arts, Philosophy, Religion, Industry and Technology, and as we have seen this is one of the key reasons why we, *Homo sapiens*, are still around while the unfortunate *Neanderthals* exist now only in the pages of history books.

Are we unique, yes, but we must remember we are also merely the very fortunate recipients of the 1st prize in nature's lucky draw. With our collective toolkit of key adaptations we won the food-hamper at evolution's village fete, but in reality any social animal endowed with the same evolutionary bag of tricks would very likely follow a very similar progressive path from tree-dwellers to Stone-age hunter gatherers to Iron-age tool makers, and ultimately to our technological developments and scientific revolution of the last several millennium.

Yet despite our clear evolutionary advantage there is still it seems a great deal of Corporate coin spent on research that would seek to add a little spice and some slightly more exotic fruit to our human "food hamper" by developing and making available technologies and practices to artificially enhance human mental and physical capabilities, and at one end of this murky spectrum is the philosophy of the *Trans-humanists*.

Trans-humanists argue that there exists within us all a duty to strive for perfection in the human condition, and that the pursuit of some sort of "Homo Perfectus" is merely the logical next step, one achieved by seeking to control our own evolution thru deliberate change (presumably we all need to look like swimsuit models, sing like nightingales, have Olympic gold medal potential in multiple sports, spend our days contemplating string theory and the meaning of life, and at the same time have the virility of a nymphomaniac). Such "deliberate change" includes research into life extension strategies, artificial intelligence, robotic prosthetics, and mind-transfer.

As an example, allegedly "reliable" sources report that research on brain and body alteration technologies has accelerated under the sponsorship of the U.S Department of Defense which is interested in the battlefield advantages they would provide to potential "Super-Soldiers". Picture a world policed by an unhealthy mixture of Robocop and Jean-Claude Van Damme and you may be close to glimpsing a possible trans-human world. To many this clearly has "nightmare just waiting to happen" written all over it, and seems closer to the science fiction world of Isaac Asimov and his like, while to others it is merely a human imperative. Regardless of individual views, and however futuristic the goals may seem, the philosophy of *transhumanism* is alive and well, and the technologies it advocates are presumably drawing nearer each day (albeit for the moment it seems behind locked doors, deep in a military bunker somewhere out in the Nevada desert).

Much more in the realms of today's mainstream science is the research into selective breeding and genetic engineering. Selective breeding (also called artificial selection) is the process by which humans breed animals and plants for particular "desirable" traits. Now selective breeding is by no means new, humans have tinkered with the hereditary blueprint of species, for thousands of years through artificial selection, most predominantly in the breeding of domesticated cattle for farming, cats and dogs for pets, horses and pigeons for racing, odd shaped goldfish, and some cross-breeding seemingly just for the cool names it created, such as the crossing of zebras with horses and donkeys to give us the wondrously sounding zorse and zonkey.

We humans have also subconsciously practiced such selective breeding in and amongst ourselves ever since we started to live in groups large enough to offer something of a choice in who we selected as our mate. Humans will naturally be drawn to the prettiest, strongest, smartest, wealthiest, most powerful members of the opposite sex, (having long flowing hair and a cool tattoo also seems to help), while those who are more physically challenged, less attractive, or are not quite the sharpest tack in the tool box, tend to find selecting (or being selected as) a mate somewhat more difficult.

But more recently however even the difficult task of securing a willing mate with the appropriately Olympic-standard genes to ensure healthy, intelligent, and attractive offspring, can be made far easier for those willing to part with the suitable sum of money as profit seeking Companies have emerged that offer human *In-vitro-fertilization* using either donor eggs or donor sperm that has been harvested from anonymous donors based upon their (supposed) superior genetic makeup. You merely turn up with your wallet and wish-list, specify the type of genetic traits you wish to merge with your own presumably already superior genes, pay the appropriate fee, and off you go with your cocktail of selected human traits ready for you to joyously bring into the world as your own.

As an interesting side-note, take time to also think for a minute about how outside of his necessary contribution to the current process of perpetuating the species, the male of our species' only other key evolutionary advantage seemingly not provided for by women (as arguably the better adapted to be the hunter-gatherer), has long since been negated with the advent of the Supermarket. Supermarkets which themselves are now just one scientific advance in fertility drugs and a corresponding product down the pharmaceutical aisle away from making the male of our species surplus to any evolutionary requirements altogether. Now that is a worry, particularly if you've just forked out for a tattoo and some hair extensions.

Genetic engineering (also called genetic modification), on the other hand is the direct manipulation of an organism's genetic blueprint using techniques that alter their genetic makeup to either remove unwanted genes (hereditary material) or that introduce new more preferred genes. It is a common belief that a fair portion of the Eastern Bloc Olympians of the 1970's had been somewhat "engineered" to perform to gold medal winning standards rather than "evolved" as such by fortuitous natural processes, and a quick review of the East German and Russian "female" shot-putt and discus throwers of the day would no doubt raise the eyebrows of today's more thorough Olympic drug-testers. Many would likely have passed the "eligibility" tests for both the men's and women's events in their chosen sports.

Since the early 1970's (at least in eastern Europe it seems) the technology for such genetic engineering has been well understood, and we humans are now (to some extent at least) able to control our own biological and evolutionary development, and thus gives us the power to modify the course of our own evolution (it's really only laws, ethical and moral concerns, and just plain fear of the potential consequences, that is stopping its general acceptance and use).

Of course the key benefit of genetic engineering is not just the chance to fill our world with blonde, blue-eyed, Olympic medal winning super-kids with IQ's that make the rest of us seem as dumb as a box of rocks, it's the potential to cure numerous genetic diseases such as cystic fibrosis by isolating and removing the culprit gene and thus removing it from the gene-pool altogether, and that of course is a very good thing.

But we can't escape the fact that we humans value beauty, strength, the ability to moon-walk like Michael Jackson and be able to fart the melody of your favourite song, above many others (a sub-conscious throw-back to our earlier Fred Flintstone days when our daily survival depended very much upon us either being, or hang out with, the athletes, super-models, cool-kids, or the brains, of the day),and genetic engineering has the potential to modify such traits on demand. For example, many pharmaceutical companies are currently spending billions of dollars trying to isolate the so called "fat gene" which if identified and successfully isolated, would likely lead to a stampede (actually, realistically more of a mass, slow-moving, waddle) of lipid-challenged individuals looking for the ultimate diet solution, and likely make billions for the patent owning company.

However if genetic engineering were to become common practice and widely available amongst our typically insecure "never too rich, never too skinny" species, we must also consider the idea that individuals who have undergone a gene altering procedure may well start to feel just a little "superior" to those left to survive with only what mother-nature gave them. Such distinctions may well lead to social issues between humans that have been "engineered" and humans that have not, and this all has a budding science fiction movie written all over it, and as we know such movies do not usually have endings that include sunsets, soft music, and everyone living happily ever after.

But even if we were to safely evolve, create, or manufacture, ourselves into some future "Homo Perfectus" superhuman species, this would of course still not in itself assure our survival as a species. The dinosaurs were hardly the perfect tool making, future explorers of space, but similarly they were not under any real threat of extinction either until an asteroid the size of Switzerland suddenly dimmed their sky and condemned them all to becoming fossil fuel overnight.

As wondrous as our universe appears to be, it remains a very unpredictable place, and just when all appears well it has a nasty habit of throwing us particularly nasty curve balls (Hurricane Katrina, the Ethiopian drought, the Haiti earthquake, the Black Death, Simon Cowell). Then of course there is the man-made disasters that often occur when we put our fate in the hands of our sometimes not completely clued-in fellow man (Napoleon's winter invasion of Russia, Lord Cardigan and the charge of the Light Brigade, the Spanish Armada's decision to sail headlong into the worst storm in a hundred years, Custer at the Little Big Horn, and the Roman town planner who decided to build Pompeii at the foot of an active volcano). As such, even finding ourselves sitting comfortably perched at the top of planet Earth's evolutionary ladder does not guarantee we will be the ones writing tomorrow's history books.

21

A Shot In the Foot

It seems that while we all sit comfortably at home on our sofas watching daytime T.V and dreaming of a lottery win, we remain thankfully oblivious to the numerous risks and unexpected surprises that may appear unannounced at anytime outside our front doors, risks that may threaten our very existence, or even that of planet Earth as a whole. As Jim Morrison of The Doors once prophetically announced, "the future's uncertain, and the end is always near", or to put it another way, if everything seems to be going well, you've obviously overlooked something.

Some such risks can actually be of our own making (global warfare, bioterrorism, global warming, reality T.V, and Boy-bands) while others are out of our control entirely (asteroid collision, super-volcano eruption, global pandemic, Darth Vader, and the Klingons). Either way our existence remains a remarkably precarious affair, regardless of how pretty we look, how intelligent we are, or how far up the evolutionary ladder we sit. The reality is that we have existed as a species for less than the blink of an eye in terms of Earth's 4.5 billion year history, and a significantly smaller period of time if compared to the almost 14 billion years of our Universe's existence, and if we wish to stick around for any real significant length of time, we will surely at some point need to face (and somehow survive) at least one event that will threaten our very existence.

It's a sobering thought that more than 99.99 percent of all organisms that have ever existed on planet Earth are now extinct. As new species evolve to fit ever changing ecological niches older species fade away, this is the natural order of things, but at least a handful of times in the last 500 million years or so up to 90 percent of all living species alive at the time have fallen foul to the deadly combination of bad timing and some catastrophic natural disaster such that they unfortunately turned out to be wearing greased trousers on a slide to oblivion, and thus vanished from the face of the Earth almost overnight.

Such catastrophes are known as "extinction events", and within Earth's history science bods acknowledge at least 5 such catastrophic incidents. The last such mass extinction happened around 65 million years ago when the dinosaurs had their untimely encounter with something significantly larger and only slightly more dense than themselves, although an extinction event approximately 250 million years ago was the deadliest, an event that caused over 90 percent of all living organisms at that time to be moved from the "currently active" to the "sadly no longer with us" list on Earth's record of species.

Generally, the impact of large asteroids or volcanic eruptions are the prime suspects as the cause of such events, where outside of the obvious devastation caused by the initial impact or eruption (a rather large dent in the ground or a nice new layer of molten rock), both would go on to eject millions of tons of debris into the atmosphere, darkening the skies for years. Starved of sunlight, plants and plant-eating creatures would quickly die, with those that had evolved to feast upon such creatures soon following suit as they started to find the larder now bare. However an extinction event that occurred around 450 million years ago involved massive glaciations that locked up much of the world's water as ice and caused sea levels to drop precipitously, and thus in the process wiped out most of the marine organisms of the day when presumably they were either frozen to death in the ice, or not physically prepared to put on snow-shoes and suddenly become land creatures.

What's important to note here is that the dinosaurs (for example) had been perfectly happy wandering around in Jurassic Park and were under no real threat from any evolutionary cuts when all of a sudden their wheels fell off as a moon-sized meteor suddenly appeared in their sky and very soon after slammed straight into them, abruptly ended their reign at the top of the food chain. As a species they had done nothing wrong, were thriving well, and outside of the obvious threat to their intended prey they weren't upsetting anyone, but they just found themselves in the wrong place at the wrong time, and so without any explanation or letter of apology from Mother Nature to Dino HQ, they were promptly evicted from Jurassic park almost overnight.

Now the dinosaurs unfortunately didn't have some Bruce Willis "save the world single-handedly" type at their disposal to save them from that giant earth bound rock, but similarly there is no real guarantee that any such hero would be around and available for gigs if another such marauding giant rock was to blip on NASA's radar screens today. It is a worry, and maybe we should all quickly compile a pre-apocalyptic bucket-list should Mr. Willis indeed prove otherwise engaged, however, we should take at least a little comfort in that the threats to the future existence of our species from such cataclysmic events as an asteroid collision, or a super-volcano eruption, although supported by precedent, are indeed quite rare, with as stated above just five such events acknowledged over the last half billion years or so (odds we could probably just about live with).

However, slightly more worryingly, history has proven that as a species we are particularly fragile when it comes to the perils of certain diseases, particularly when the list of symptoms for such a disease includes imminent death, and the disease itself decides to go global. In the last few hundred years alone we have faced the joys of such global pandemics on several occasions, each of which has been responsible for reducing our total number by more than just a few decimal places.

Dating back to the 1500's, smallpox has since ravaged entire populations, and caused the death of nearly 90 million Native Americans alone when European settlers rather unfortunately brought it over (along with their "iron-horses", guns, whiskey, and "civilization") as an unwanted gift to the Americas. The Spanish Flu of 1918 is widely known as one of the most deadly pandemics in history, and though it only lasted for a year, it was responsible for the deaths of 50 to 100 million people, while the Black Death (which for the colorblind is also known as the Bubonic Plague), is considered to be the first true pandemic, cutting populations in half throughout parts of the world like Asia and Europe in the fourteenth century. Even as recently as the 1980's, the still incurable disease AIDS has itself claimed more than 25 million lives. It will only take one deadly new strain of something particularly nasty to hop on some passenger planes and quickly distribute itself around the world before our medical boffins can develop a cure, to potentially end our glorious evolutionary run.

Other often cited potential sources of our species demise may also be a little way off into the future (hopefully) given their very nature. Cosmology boffins with time on their hands have calculated that the Andromeda Galaxy is seemingly on a collision course with our own galaxy, the Milky Way. It appears that Andromeda is approaching at an average speed of about 110 kilometers per second, a frighteningly speedy pace you would have thought, but before you rush off to hand in your resignation and start making your way through your "bucket list", note that given the enormity of galactic distances such a collision is not predicted to occur for at least another four billion years even at those break-neck speeds.

Although evidence of alien life on other planets has never been documented, scientists do believe that the existence of such extraterrestrial life is still very likely. However the belief held in some quarters that there is an imminent threat of an Alien invasion of Earth to either exterminate and supplant human life, enslave it under a colonial system, harvest humans for food, steal the planet's resources, or destroy the planet altogether to make way for some newly planned inter-galactic highway, does seem a rather unlikely end to our species.

And something worth bearing in mind next time you start to swear at your non-responsive laptop is the suggestion that at some point in the future "artificial intelligence" computers may well "evolve" to become "super-intelligent" and ultimately "self-aware", and may seek to foil any attempt that threatened to prevent it from achieving its goals, ultimately wiping out any other challenging rival intellect such as mankind.

Another interesting possibility for the brains-trust bods at Science HQ, but a rather more worrying possibility to the rest of us, is the threat from the Sun's solar flares. Solar flares are a consequence of the sunspots we sometimes see appear on the Sun's visible surface (the photosphere) as dark patches, and which are just areas of gas that are cooler than the gas around them. Sometimes these areas of "cooler" temperature erupt in a huge release of energy which is known as a solar flare and which we on Earth witness as a plume of light erupting out at the sun's edge. These flares throw out dangerous and powerful radiation into space (an average solar flare will typically release the same energy as a billion or so hydrogen bombs, and thus fire out into space billions of tons of associated deadly radiation) and as they penetrate the earth's magnetic field they are thankfully safely directed to the earth's poles (all due to some clever magnetic jiggery-pokery beyond the scope of my intelligence, and thus also beyond the scope of this book) where they merely create the picturesque atmospheric lightshows called aurorae (the Northern Lights), and sometimes the associated static will even cause temporary blackouts in things like radio communications.

The worst solar flare event, at least in recorded history, was a so called solar super-storm which occurred in 1859, and known as the "Carrington Event" after the British astronomer Richard Carrington who first spotted it. It lit up the skies across the world with stunning aurorae, and the associated electrical surges induced in the Earth's magnetic field were reportedly sufficient to power the telegraph systems without batteries, with some excitable telegraph operators even reporting sparks flying from their equipment. It was all very dramatic but, outside of some interference with Victorian communications, there were no long-term effects. Clearly a similar sized event today would at the very least have far greater consequences given our modern day dependency on electrical and electronic gadgets.

Now what keeps the solar flare boffins up at night is the worrying possibility that one day the Sun for some reason may decide to belch out a giant solar flare hundreds of times larger than the typical flare we see thrown out on a regular and happily somewhat inconsequential basis, something that is a "distinct scientific possibility" these happy folks at Sun-Watchers HQ tell us. The resulting wave of deadly radiation would likely engulf the Earth. On the positive side it would create the most spectacular lightshow ever seen, but on the minus side it would likely be the last thing any of us actually saw.

However, as unlikely as some doomsday scenarios may seem, it may be worth noting for future reference that Sir Isaac Newton (not a man who was wrong very often) surmised that the end of the world would happen no earlier than 2060, although the earlier threat of the world's end as predicted by the Mayan "Long Count" calendar, thankfully passed on December $21^{st,}$ 2012 with little incident outside of a few rash job resignations and as it turned out some unnecessary personal confessions.

Now before we all rush off to start our own "Coping with Doomsday" support groups, in reality it seems that when it comes to any real immediate threat to our current lofty pole-position in nature's evolutionary race we do edge much closer to the realms of sphincter-tightening odds when it comes to the very events over which we do have the most control, primarily because we are the very originators of their existence.

Although we do appear to be our little planet's supreme evolutionary achievement (at least in our own clearly unbiased opinion), we are also in many respects our planet's worst enemy, we are very much the Jekyll and Hyde species of our little ball of rock, and as such, in pursuit of our own gain we are a singularly destructive breed. Outside of our willingness to wipe-out whole generations of our own kind for the sake of land ownership, an unfortunate misunderstanding, or more often than not over the name of whose God should appear on the Universe's letterhead, we have also been directly responsible for the extinction of hundreds, if not thousands, of other species, just ask the Dodo, the sea cow, or the Tasmanian tiger (well actually, no, we can't now can we). We also cut down areas of rainforest the size of small countries every day, fill the atmosphere with dangerous pollutants, over fish the seas, pour toxic chemicals into the rivers and lakes, cover large areas of farmland with harmful pesticides, and all with seemingly complete disregard for what affect we may be having on the planet and its millions of other species, let alone the countless sleepless nights it causes for the likes of poor Sting and Sir Bob Geldof.

One such example has given rise to a crisis so large that science bods have given it its own name, Global Warming, and refers to the abnormal warming of the planet caused by human technology, which seems to have had the rather unfortunate consequence of ensuring that adverse climate change and related disasters are on the rise. Around 70 percent of recorded disasters are now apparently climate related, significantly up by around 50 percent from two decades ago (we should note here for the record that statisticians report that 82.23% of all statistics are about as accurate as a drunk man taking a pee), while in the last decade alone it seems that around 2.4 billion people have been affected by climate related disasters.

Destructive sudden heavy rains, typhoons, intense tropical storms, repeated flooding and droughts, are all it seems likely to increase, and sea levels may even rise to levels that will completely inundate certain low lying coastal areas to such an extent that Wellington boots would become mandatory footwear in London and New York, those wishing to continue to live in places like Miami, Mumbai, or Osaka, would probably first need to grow fins, and the places on the world map where countries such as Bangladesh and the Netherlands currently sit would likely need to be coloured blue altogether.

Global warming is caused when the level of carbon dioxide (CO_2 for the budding chemists amongst us) increases in the atmosphere to a point where it starts to absorb more and more of the sun's heat that has been radiated back up from the Earth's surface, and thus in turn "reflects" back more heat, that then further warms the Earth. In simple terms, the more CO_2 that is present in the atmosphere the more heat will be absorbed and thus sent back to the Earth's surface. It is similar to the process that turns the interior of your parked car on a hot summer's day into a furnace hot enough to smelt iron, or that allows keen gardener's to grow prize winning marrows in greenhouses. Unfortunately, since the Industrial Revolution, mankind has been pouring huge amounts of the offending gas into the atmosphere by our ever growing use of fossil fuels.

Now I'm not Sting or Sir Bob Geldof, I don't hug trees, I don't lay in front of bulldozers to prevent areas of natural beauty from being flattened for a new parking lot, I'm not a vegan or a member of Greenpeace, but even I can see from my distinctly less-than-green perspective, that global warming is something we may just want to start to pay a little attention to.

The rather embarrassing facts are that the increased CO_2 in the atmosphere due to our excessive burning of fossil fuels means that we are now seeing a marked increase in the earth's temperature, and thus we are starting to experience all that that implies in terms of climate change and associated natural disasters of ever increasing magnitude. Unchecked we may likely be the first species to be inadvertent architects of our own extinction event. Quite the shot in the foot, and quite impressive for a species supposedly representing the peak of the evolutionary process (dolphins are probably happily clicking "I told you so" to each other as we speak).

And as if this little human contribution to the destabilization of the Earth's fragile balance doesn't slap enough egg on our collective faces, consider this additional twist provided by mankind in the Global warming story. In Earth's natural state CO_2 levels are actually kept in balance by the activity of trees and plant life. Trees and plants use a complex process known as "photosynthesis" to live, essentially using energy from the sun to transform carbon dioxide and water into glucose for food, and as a welcome by-product of the process they release oxygen back into the atmosphere. Thus the trees of the rain forests effectively act as a huge CO_2 filter for the Earth, removing the harmful gas and replacing it with life-giving oxygen, in other words, while we inhale oxygen and exhale carbon dioxide, trees inhale carbon dioxide and exhale oxygen, which is a very welcome trick for all of us oxygen breathers and a convenient temperature thermostat for the planet as a whole.

But as we humans burn more and more of the harmful CO_2 releasing fossil fuels, the planet's CO_2 levels have shifted so far out of balance that we are now emitting way more of the offending gas into the atmosphere than the poor overworked plants and trees can handle, and here's the twist, in an entirely separate act of self-serving destruction (and despite no end of complaints and letters from Sting, vegetarians, tree-huggers, and various people in knitted cardigans) mankind continues to lay waste to much of the world's rainforests thru a process of deforestation. Thus, not only are we filling the atmosphere with more and more CO_2 thru our fossil fuel emissions, but we are at the same time gradually removing the very tool nature has kindly put in place to help keep the levels of CO_2 in check in the first place. By filling the atmosphere with unwanted additional CO_2 and then also systematically removing vast areas of rainforest we have effectively produced something of a full-scale self-inflicted double whammy. No wonder Sting's upset.

And it's not just thru Global warming and deforestation that we may see mankind shoot itself in the evolutionary foot. Remember it was not too long ago that we were on the brink of nuclear war as super-powers played a game of brinkmanship with one another, and a nuclear Armageddon was a real possibility. Today we face no less of a threat from chemical weapons and so called weapons of mass destruction. It would only take an over inquisitive cleaner at the White House to ask themselves "I wonder what this red button does?", or a slightly unstable megalomaniac whose mommy didn't show him enough love as a child, who has enough money, power, and resources, to one day get out of bed on the wrong side, and we could all be history by dinnertime.

In the last 200 years alone we have fought two World wars involving most of the nations of the world and which were collectively responsible for the deaths of almost 100 million people, either through direct conflict or indirectly through war induced epidemics, diseases, or famine. Within the same short timeframe we can add the Napoleonic wars, the Vietnam war, the Crimean war, several acts of mass genocide, and countless civil wars, all clearly indicating that as a species we tend to resolve our differences with weapons rather than words. It also seems that we are the only species on Earth that behaves this way, there have (so far at least) been no recorded instances in history of herds of cows taking up arms against their fellow cows in a dispute over grazing rights, or great flocks of pigeons gathering over Europe in preparation for an invasion of Poland.

And something else for us to consider next time you look to "super-size" your meal or listen to the last slice of cake quietly calling you from the back of the fridge, it appears that there is also the looming threat of human overpopulation as it seems that we are increasing our global headcount at a rate that the rabbits down at your local warren would be proud of. Scientists believe that we are all too quickly heading towards an agricultural crisis where food production will no longer be able to keep pace with worldwide population growth (It seems the "catastrophe" predicted by Thomas Malthus that so influenced Charles Darwin still holds true some 150 years later). They predict that such an agricultural crisis will likely begin to impact us around 2020, but before you all rush out to stockpile your cellars with tins of beans and long-life milk, it seems it will not become truly critical until around 2050, but that ultimately we will see spiraling food prices, food rationing, associated civil unrest, and massive starvation on a global level as demand for food quickly outgrows supply.

And as an interesting side note to this looming threat to our expected three square meals a day, we should also be aware that the male of our species clearly prioritises food above just about everything else, except maybe sex, although the ability to quickly and easily satisfy a need to eat generally means that food wins out over sex in most cases, as sex (at least in its traditional sense) requires the difficult additional task of first finding a willing partner to complete the act. This is one of the main reasons why our high streets are full of fast food stores and not florists.

And so as a species that presumably wants to be around for many years to come, we have the double advantage of possessing unique physical adaptations and the scientific and technological knowledge to allow us to plot our own destiny. But as an inhabitant of planet Earth we also must face the vagaries of the universe (and our own stupidity) which may see us only a giant asteroid, super-volcano eruption, one too many fossil-fuel driven factories, or someone inadvertently pressing the wrong button away from an entry in the obituary section of The Daily Universe.

Now Sting and his buddies maybe a little more sensitive to all of this than most of us, but the lesson seems as clear as a Valentine's day rejection, we must all learn to accept our responsibility in ensuring that as a species we act not in a manner driven only by profit and self-serving gain (if at all), but in a way that acknowledges that to survive as a species for any length of time we must also act in a manner that protects the fragile balance of our planet. The alternative may well be that we rather embarrassingly go down in history as the first species to understand exactly what the potential devastating consequences of our actions would be, but went ahead and did it anyway (you can almost hear Westminster Abbey rumbling to the sound of Darwin turning in his grave).

It's enough to put you off your beer and nibbles, but it's a thought that maybe we should also embrace with a degree of optimism. There has been a significant amount of good fortune involved in getting us to where we are today, and we are (hopefully) still only at the beginning of our journey. There is a universe of possibilities open to us, but we must also embrace that future with an understanding that no matter how fortunate our own evolutionary path, our existence is no more important than that of any other creatures. We have a responsibility to respect not only each and every living organism we share this planet with, but also a responsibility to respect the planet itself.

What we do and where we go next is up to us. The good news is that the choice is entirely our own, the bad news is that recent history has shown us we do not always act with that responsibility in mind.

22

Just an Alien Experiment after all?

And so, as the trillions of tiny cells that make up the unique being that is you all seamlessly come together to enable you to turn over the last few pages of this book, we have arrived at the end of our brief tale of human evolution. It's been a long journey, one that started with quite literally nothing, which somehow exploded into absolutely everything, and then some 13.5 billion years later through the miracle of evolution and the good grace of Lady Luck, a little bit of which turned into you and me.

It has been an ambitious tale too, we've seen a universe built from scratch that quickly filled with all sorts of wondrous objects, we've seen life created on a ball-shaped, rather rocky, and very average example of one of those objects, which evolution then went on to evolve into (amongst many other curious creatures) a quite extraordinary and very fortunate biped that itself went on to provide the Universe with such glorious wonders as sliced bread.

I am grateful that you felt the journey engaging enough to get to this point, and frankly a little surprised, likely less surprised than the Captain of the Titanic when suddenly confronted by that not inconsequential iceberg, but nonetheless still surprised for this has not been an easy tale. The journey was a complex one, which is why we had to wait for boffins with brains large enough to figure most of this stuff out to come along to help explain it all for us.

Each tiny little piece of the puzzle has been put together by a constant act of intellectual yoga and advanced mental gymnastics by our brightest minds, and even now, as the previous pages have shown, we don't really have all the answers we just have hypotheses that we can try to string together into a theory that tries to explain it all. But as we have also seen, particularly in the science of human evolution, even those boffins with the watermelon-sized brains often can't even agree on the hypotheses in the first place, let alone how to string them all together, thus making the whole field really not much more than an exercise in well informed guesswork.

We've also covered a great deal of ground, mostly riding on the shoulders of those great minds, certainly minds greater than I would like to go up against in any quiz night down the local pub. We looked at how life could have possibly crawled out of a primordial toxic soup, followed Charles Darwin on his path to evolutionary theory, and then witnessed such evolution at work as an ape-like creature decided to climb down from the trees to discover life on two legs, fire, a smart way to move objects more efficiently, build tools, and ultimately to invent some numbers and clever ways to use them to try and explain all of this to the likes of you and me.

However, there has been no pretence within the covers of this book at any attempt to formally teach anthropology, biology, or evolution, this is what schools were invented for. And for those who chose to spend their schooldays either smoking behind the gym, playing video games, or bullying smaller kids, rather than paying any attention to the failed scientist turned educator standing at the front of the class, there are night-schools, "Science for Dummies" books, and doctor's waiting rooms full of National Geographic magazines, to provide the required second chance at some academic enlightenment. There will certainly be no test for you to sit once you shortly turn the last page of this book, and I would strongly advise that you do not even contemplate sitting any test at all armed only with the half-baked facts presented within the covers of this book.

No, as I stated right at the outset of our journey, this was to be just a story, a possible version of events that starts on day-one of the universe and through a series of seemingly random events takes us through the several million year evolution of ape, to ape-man, to man, from *Ardipithecus* to *Australopithecine* to *Homo sapiens*, and finally resulting in trillions of cells somehow assembling themselves in a complex and very obliging manner, working together like a well-oiled piece of German engineering to create the very unique entity that is you, a current day living example of genus *Homo*, species *sapiens*.

The purpose of each of the preceding pages was less about teaching science and more about trying to inspire a respect and a sense of wonder for the universe, the miracle of the life that lives within it, and the fortuitous evolutionary path that ultimately led to our species becoming the intellectual masters of our planet. And although to many of us anything more than just a cursory understanding of the chemistry of life's creation, or the mechanics of evolution, are just as beyond us as a date with a swimsuit model or an ability to become a Superhero just by wearing our underpants on the outside, this anyway is not the path to any true understanding of what is really important.

Unless your surname is Hawking, Einstein, or Newton, what is important is to simply understand that from the mechanics of the smallest cell, to evolutions twists and turns, in fact to the workings of the universe as a whole, there is no real guiding purpose to it all, no aim, no goal, no "bigger picture", no Captain of the good ship "The S.S. Universe", yet somehow the incredible majestic dance of the galaxies and the stars that we see around us, the wonders of the multitude of life we see on our planet, and the random forces of evolution that has created them, the miracle that is you and me, magically all just happens, and it all just works. This is the true miracle of life and wonder of the universe, that the random, mechanical, unthinking nature of every single atom in the universe have all, without any direction or purpose, somehow miraculously found a natural collective harmony that makes life, makes you and me, makes everything else in the universe, all possible.

Similarly, I hope the pages of this book have also helped teach an understanding of just how incredibly lucky we as a species are to be here. If the preceding pages have highlighted anything it is that it is clearly not easy surviving for any length of time on planet Earth, as the 99.99% of the all species that have ever lived on its surface and who now find themselves consigned to the "sadly no longer with us" list, would testify too. By the same token, if we want to stay here for any further length of time it seems clear we will need a great deal more of that good fortune and importantly exercise a great deal more common sense than currently being displayed on our part. After all, we are the result of 4 billion years' worth of evolutionary progress, we really need to start to act like it.

The process of evolution we have seen is primarily a random one, and to start with a simple single-celled bag of chemicals and then over the course of 3.8 billion years to somehow arrive at the upright, free-thinking, intellectually superior species (in theory at least) that is you and me, has required an extraordinary series of fortuitous genetic mutations, evolutionary twists, and some down-right lucky breaks.

Consider for instance that without the vagaries of our planet's climate our early ancestors may well have just stayed up in the trees (which is presumably where you and I would still be today) and never ventured to step down onto the open savannahs of Africa and looked to walk around on two legs. Or a few million years later driven by the same climatic challenges, survive only by learning to adapt by fashioning crude tools out of stone, or the remarkable slice of evolutionary "luck" that somehow prompted the brains of those early ancestors to suddenly grow such that they would find ways to build ever more complex and useful tools, expand their territory to cover the 4 corners of the planet, and ultimately walk on the moon.

And for you as a living breathing individual example of our fortuitous species alive today in the 21st century, you are doubly lucky. Because for you to be here today sat reading the pages of this book, means that you are part of a singularly unique hereditary line of individuals tracing back through your own unbroken family tree that spans 3.8 billion years, and thus one that has somehow survived generation after generation under the constant threat of predators, disease, famine, war, unfortunate genetic mutations, and somehow survived countless meteor hits and extinction events, all without a single break in your ancestral line.

Your father, your father's father, and back through each preceding generation, through each evolutionary incarnation of your family tree from single-celled organism to fully fledged *Homo sapiens*, via brief periods as a fish, an amphibian, a reptile, and an ape, through almost 4 billion years of evolution, each and every one of your own unique family tree's thousands upon thousands of previous generations have all somehow managed to survive everything evolution and the universe could throw at them. And remember, all this while at the same time also managing to successfully complete the necessary trick of not only finding a suitable mate (not an easy task in itself as any spotty-faced teenager will tell you) but also ensuring a healthy hereditary line to the next, and all resulting in your illustrious presence on Earth today.

It would only have taken just one single ancestor of yours somewhere within those thousands of generations to have strayed too close to a T-Rex, stood and fought when they should have run, been so singularly unattractive as to not have found a mate, or stayed in bed the day that everyone else decided to step down from the trees, and sadly you as an individual would just not be.

And now having arrived, and as an individual hopefully finding yourself lucky enough steer a course through life's unfortunate curve-balls of sickness and disease, and also managing the increasingly more difficult feat of ensuring you are not in the wrong place at the wrong time when any one of a growing number of possible natural and man-made disasters strike, you can expect to live a life adding up to about 80 years in length. That's around 700,800 hours or roughly 2.5 billion seconds. Given that planet Earth has been around for about 4.5 billion years, and the Universe for approximately 13.5 billion years, you can see that we are gifted only an incredibly fleeting period of time with which to make the most of our own existence here on this Earth. Remember these numbers next time you find yourself vegetating on the couch still in your pyjamas at 3 o'clock in the afternoon because you can't even muster the enthusiasm to get vertical let alone dressed and active.

And while doing so also remember that regardless of your race, colour, creed, sex, physical appearance, brain size, shoe size, breast size, or beliefs, we are all descendants of the exact same ape-like ancestors, we are all ultimately related to one another and as such are all equals. Remember, we only need to go back a mere 7 million years or so and we weren't "ape-like" at all, we were fully "ape". Even today, despite our ability to fly rockets to the moon, create achingly beautiful works of art and music, and feast on bread that has miraculously been pre-sliced, we are still almost 99% genetically identical to chimpanzees.

Go back even further to that first emergent life that one day some 3.8 billion years ago popped its head out of the primordial soup, liked what it saw and decided it was worth sticking around for a while, and you will see the single common ancestor of every living creature that has ever graced our rocky planet. As such, we are all ultimately related to every living creature we see around us, and in turn are all a part of the same random but joyously harmonious dance that is the Universe around us.

Understand this and the respect it implies we owe not just to one another but to every other living creature and the giant ball of rock we all share and call home, then act accordingly for every one of the 2.5 billion seconds you will hopefully be counted as a living sentient being on this planet, and our species may well have half a shot at living beyond the next millennium.

Of course there is still the off chance that we are indeed really all just part of an ambitious alien laboratory experiment, in which case ignore everything that's been said so far, go ahead and smoke, drink, covet your neighbor's wife and that expensive sports car, chop down trees and burn fossil fuel with reckless abandon, as we're all just a galactic funding cut away from having the plug pulled on the Universe anyway!

++++++++++++++

www.ingramcontent.com/pod-product-compliance
Lightning Source LLC
Chambersburg PA
CBHW060844170526
45158CB00001B/229